Trigonometry
Lessons That Speak To You

by
Fred Truong

Table of Contents

Preface

Dear Students:

Are you worried that your math grades will keep you from attending the college of your dream?

Do you feel like you want to major in Science, Engineering, Architecture, Economic, Business or Finance, but afraid maybe it has too much math?

Are you puzzled by math concepts such as sine, cosine, tangent, cotangent, secant, cosecant, radian angles, trigonometric graphs, trigonometric equations, etc.?

Do you wish to have access to study materials that can provide clear explanations of how to solve math problems?

Hi, my name is Fred Truong. I am a professional math tutor for over twenty years. If you answered "yes" to any one of the above questions, and you:

1) are going to take Precalculus/Trigonometry and wish to get ahead, or

2) are currently taking the class and need extra help or want a deeper level of understanding, or

3) had taken the class, but wish to review the materials for AP Calculus or the subject test (SAT II),

then this book is for you. I wrote it the way I would speak it, and the result was, every lesson contains the clearest explanation possible. It was as if I were mapping my thoughts from my brain to yours. Have you wondered why almost nobody read his or her math textbooks from school? The reason is simple; those books are unreadable. The authors do not speak to you, the audiences; they often use complicated languages just to explain things on the surface level. You will find that my lessons are different. I use simple language, but yet go deeper into every math topic. So without further ado, let's get started.

> ## Did you know?
>
> To multiply a two-digit number by **11**, you can simply add the two digits and insert the sum in the middle of the two digits. For example, 34 x 11 = 3**7**4 because 3+4 = 7.
>
> If the sum is more than 10, start from the right and "carry" over. For example, 58 x 11. First write down **8**; then **3** because 5 + 8 = 13. Finally, carry the 1 over to the 5 and write down **6**. The answer is **638**.
>
> Now how about multiplying a large number by 11? For example, 3,215,418 x 11. Start from right to left, write down **8**, then 8 + 1 = **9**, then 1 + 4 = **5** and so on. The answer is **35,369,598**.

Lesson 1: The Six Trigonometric Functions

As you will see, to study Trigonometry is to learn how to manipulate the six trigonometric functions: Sine, cosine, tangent, cotangent, secant and cosecant.

So, what exactly are they? Well, you may have a basic idea of, or heard about what sine, cosine, and tangent are from your previous math classes, but we will explore them more extensively here.

Let's begin by talking about angles.

You would need two sides to form an angle.

The first side is called the **initial side**.

If you put another side on top of the initial side, then you have a zero degree angle.

Now imagine if you rotate the second side counterclockwise to a new and final position, you have just created a positive angle with the second side as your **terminal side**. (I will talk more about negative angle in the next lesson).

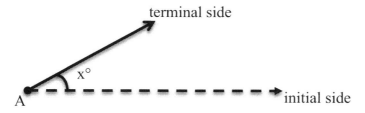

Suppose you stand at a random point B on the terminal side, and walk "straight down" to point C of the initial side, you create right △ABC as shown below.

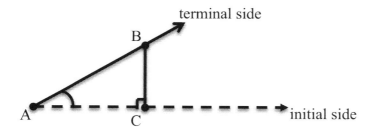

If you stand at another random point D on the terminal side, and walk "straight down" to point E of the initial side, you create another right triangle as shown.

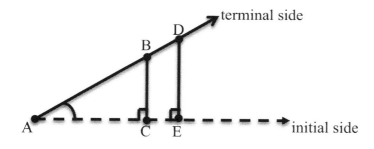

Recall from Geometry that if the measures of two angles of a triangle equal to the the measures of the corresponding two angles of another triangle, then the triangles are **similar**.

Since △ABC and △ADE share ∠A, and their right angles are equal, this means △ABC is similar to △ADE. As a result, their corresponding sides are proportional.

So, for example, $\frac{BC}{AB} = \frac{DE}{AD}$, or in other words, the ratio of the opposite side of $\angle A$ to the hypotenuse is a constant. Mathematicians named this ratio **sine** (abbreviated as **sin**).

So, by definition

$$\boldsymbol{sin\,\theta = \frac{opposite}{hypotenuse}}$$

Note: *The θ symbol is read as "theta." In Trig., you will often see people use the Greek alphabet such as alpha (α), beta (β), gamma (γ), or theta (θ) to name angles.*

Similarly, $\frac{AC}{AB} = \frac{AE}{AD}$, or in other words, the ratio of the adjacent side of $\angle A$ to the hypotenuse is a constant. This ratio is known as **cosine** (abbreviated as **cos**).

Here are the definitions of all six possible ratios:

1) $sin\theta = \frac{opposite}{hypotenuse}$ 4) $csc\theta = \frac{hypotenuse}{opposite}$

2) $cos\theta = \frac{adjacent}{hypotenuse}$ 5) $sec\theta = \frac{hypotenuse}{adjacent}$

3) $tan\theta = \frac{opposite}{adjacent}$ 6) $cot\theta = \frac{adjacent}{opposite}$

It is important that you memorize these definitions as soon as possible. Notice that the last three functions are just reciprocals of the first three. You will probably hear your teacher use SOH CAH TOA to help you remember sin, cos and tan as follow:

Sin	**Cos**	**Tan**
Opp	Adj	Opp
Hyp	Hyp	Adj

Now sometimes students forget whether the reciprocal of sin is sec or csc.

Just use the following trick:

"s" is the reciprocal of "c"

"c" is the reciprocal of "s"

Meaning: sin goes with csc

cos goes with sec

So as you can see, even though the names of these functions may sound like mumbo jumbo, it is nothing more than just ratios. As you continue to study Trigonometry, it is important to always keep this in mind. This way when you see something like $\sin\theta = 2$, you should know that it does not make sense. There is no angle in which the ratio of the opposite side to the hypotenuse is more than 1. The opposite side is always less than the hypotenuse. The value of $\sin\theta$ is always less than 1 for all $\theta < 90°$. In later lessons, we will talk about trigonometric functions beyond $90°$, then $\sin\theta$ could equal 1.

Now, let's take a look at some examples.

Example 1:   For a fixed angle measure such as $50°$, what is the ratio of the opposite side divided by the hypotenuse?

Solution: You can answer this question by entering $\sin50°$ into your scientific calculator. You will get $\sin50° \approx 0.766$. This means if you draw a $50°$ angle with the length of the hypotenuse equals 1 unit, and you measure the opposite side, the length will be 0.766 unit. Here is the diagram.

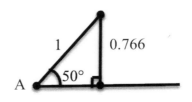

Example 2:   Given m∠A = 50° and the hypotenuse = 2 as shown below, what are the lengths of the opposite and adjacent sides of ∠A?

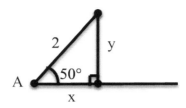

Solution: We know $\sin \angle A = \dfrac{opposite}{hypotenuse}$

So, $\sin 50° = \dfrac{y}{2} \Rightarrow y = 2\sin 50°$

$\Rightarrow y = 2×0.766 = \boxed{1.532}$

Also, $\cos \angle A = \dfrac{adjacent}{hypotenuse}$

So, $\cos 50° = \dfrac{x}{2} \Rightarrow x = 2\cos 50°$

$\Rightarrow x = 2×0.643 = \boxed{1.286}$

So, how is using Trigonometry different from the Pythagorean Theorem?

Using Trigonometry, you can solve for the whole triangle if you only know one of the acute angles and one side, or if you know two sides.

Using the Pythagorean Theorem, you must know two sides just to solve for the third side.

***Example 3*:**   Find the exact values of the six trigonometric functions for $\angle B$.

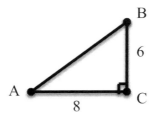

Solution: Using the Pythagorean Theorem, we have

$$6^2 + 8^2 = (AB)^2$$

$$36 + 64 = (AB)^2$$

$$100 = (AB)^2$$

$$10 = AB$$

$$\sin \angle B = \frac{8}{10} = \frac{4}{5} \qquad\qquad \csc \angle B = \frac{5}{4}$$

$$\cos \angle B = \frac{6}{10} = \frac{3}{5} \qquad\qquad \sec \angle B = \frac{5}{3}$$

$$\tan \angle B = \frac{8}{6} = \frac{4}{3} \qquad\qquad \cot \angle B = \frac{3}{4}$$

Ok that's it! Congratulation, you've just finished the first lesson. It's time for some practice problems. However, if necessary, please re-read this lesson or any future lessons at least one more time, to make sure you understand everything before you move on. Also, reading math is different from reading an English novel. Take your time and read slowly. Let important mathematical concept sink in. In other words, **keep wonder and ponder**! See you next lesson.

Practice 1

Find the exact value of the six trigonometric functions of ∠X.

1)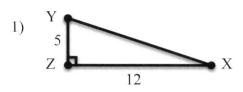

2)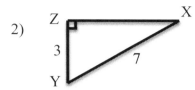

3)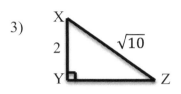

4)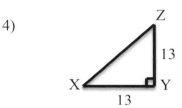

In problems 5-8, use the given information to solve the triangle shown below.

5)   m∠A = 75°, AC = 20

6)   m∠C = 40°, AB = 8

7)   m∠A = 35°, BC = 11

8)   m∠C = 19°, BC = 10

In problems 9-12, use the given information to find the area of the triangle shown.

9)   Given XY = 10, m∠Z = 70°

10)   Given XZ = 6, m∠Z = 25°

11)   Given YZ = 15, m∠Y = 34°

12)   Given XY = 2, m∠Y = 40°

In problems 13-16, use a calculator to find each value to the nearest hundredth.

13)   sec 80°   14)   csc 33°

15)   cot 100°   16) $(sin110°)^2 + (cos110°)^2$

17)   Find the area of the triangle at the right

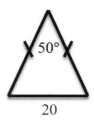

In problems 18-20, find the missing value to the nearest hundredth.

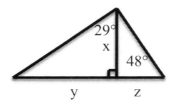

18)   Given x = 11, find z.

19)   Given z = 5, find y.

20)   Given y = 30, find z.

Bonus: Find the area of the rectangle.

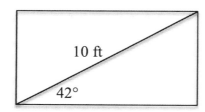

Did you know?

*To square a two-digit number ending in 5, you can simply multiply the first digit by one plus itself and insert 25 at the end. For example, $35^2 = 1225$ because the first digit is **3**; one plus itself is **4**, and $3 \times 4 = 12$. Insert 25 after 12 gives 1225.*

*Here is another example. What is 85^2? Since $8 \times 9 = 72$, the answer is **72**25.*

How about $\left(\dfrac{13}{2}\right)^2$? Since 13 divided by 2 is 6.5, $6.5^2 = 42.25$ because $65^2 = 4225$. You can do this even with a three-digit number.

For example, $115^2 = 13225$ because $11 \times 12 = 132$. (Remember the "eleven trick"?)

Lesson 2: Radian Angles

Before I go any further with Trigonometric functions, I want to show you another unit of angle measurement. Also, this time I will rotate the angle around a circle; this way it can go beyond 90, 180 or even 360 degree.

As before, if you put the terminal side on top of the initial side, you have an angle equals to zero degree or zero radian.

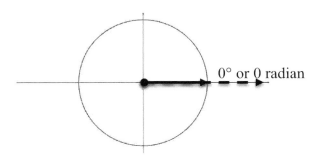

Now, what is one degree? A lot of students thought they knew what a degree is, until I ask them to give me a definition.

You form a 1 degree angle when you move the terminal side $\frac{1}{360}$ of the length of the circumference.

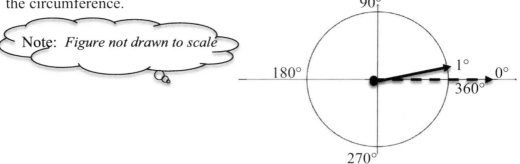

Note: *Figure not drawn to scale*

In other words, you can rotate the terminal side 360 times before it goes back to the initial position. For example, if you only turn the terminal side 180 times, it will land on the half circle; 90 times will take you to the quarter circle and so on...

Next, what is 1 radian?

On radian is an angle formed when you move the terminal side a distance equal to 1 radius of the circle. (Hence the name, 1 "radian").

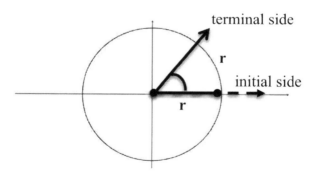

Since the circumference of a circle is $2\pi r$, which is about $2\times3.14\times r = 6.28r$, there are about 6.28 radian in a circle. That means if you move the terminal side about 6.28 times the length of the radius, you will go back to the initial position. If you move 3.14 times, you will land on the half circle. (Make sure you understand this point before moving on).

So, to convert between radian and degree, you can use the following fact:

$$2\pi \text{ radian} = 360 \text{ degree, or equivalently}$$

$$\boxed{\pi \text{ radian} = 180 \text{ degree}}$$

Example 1: Convert from radians to degrees.

a) 1 radian b) $\frac{\pi}{6}$ radian c) $\frac{\pi}{4}$ radian

Solution: a) 1 radian = 1 rad $\times \dfrac{180 \text{ degree}}{\pi \text{ rad}} = \dfrac{180°}{3.14} = 57.3°$

b) $\frac{\pi}{6}$ radian $= \frac{\pi}{6}$ rad $\times \dfrac{180 \text{ degree}}{\pi \text{ rad}} = \dfrac{180°}{6} = 30°$

c) $\frac{\pi}{4}$ radian $= \frac{\pi}{4}$ rad $\times \dfrac{180 \text{ degree}}{\pi \text{ rad}} = \dfrac{180°}{4} = 45°$

Example 2: Convert from degrees to radians.

a) 60° b) 90° c) 135°

Solution: a) $60° = 60° \times \dfrac{\pi \, rad}{180°} = \dfrac{\pi}{3} rad$

b) $90° = 90° \times \dfrac{\pi \, rad}{180°} = \dfrac{\pi}{2} rad$

c) $135° = 135° \times \dfrac{\pi \, rad}{180°} = \dfrac{3\pi}{4} rad$

Note that you can use the idea of "multiples" to do example 2c.

Since $135° = 3 \times 45°$, this means $135° = 3 \times \frac{\pi}{4} rad = \frac{3\pi}{4} rad.$

Similarly, $150° = 5 \times 30° = 5 \times \frac{\pi}{6} rad = \frac{5\pi}{6} rad.$

You can also go backward; since $\frac{\pi}{2} = 90°, \frac{3\pi}{2} = 3 \times 90° = 270°.$

Below are some of the common angles along with its positions on the circle.

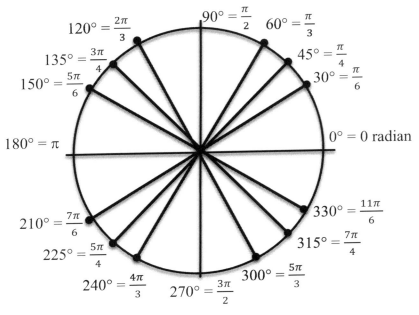

To figure out which quadrant an angle is in, it is as easy as counting. Just remember a circle has 2π radian. Half a circle is 1π; half of that is $\frac{\pi}{2}$ and so on.

You can count by $\frac{\pi}{2}$

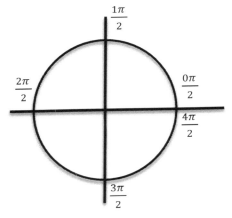

You can count by $\frac{\pi}{4}$

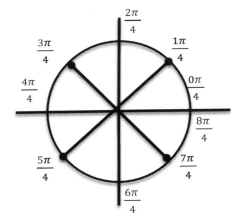

You can also count by 30° or $\frac{\pi}{6}$. Since each quadrant is 90°, you will have three sections in each quadrant.

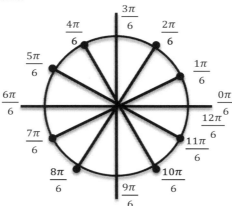

Now, you can rotate the terminal side beyond 360°. For example, if you turn 1 full revolution plus 1°, you have 361°. That means, 1° is the same as 361°. Similarly, 30° is the same as 390°. You just need to add or subtract 360 to get equivalent angles. So,

$$\boxed{1000°} = 1000 - 360 = \boxed{640°} = 640 - 360 = \boxed{280°}$$

If you keep subtracting, you even get negative angles. In this case, you just rotate clockwise instead. So, −30° is in the fourth quadrant, which is also equals 330°.

In radian, it is $\frac{-\pi}{6} = \frac{-\pi}{6} + 2\pi = \frac{11\pi}{6}$

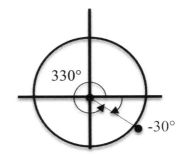

In fact, −30° and 330° are called **coterminal** angles, because they have the same terminal side.

Example 3:   Find 2 positive and 2 negative angles that are coterminal to 200°.

Solution: $200° = 200 + 360 = 560°$

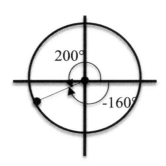

$200° = 200 + 720 = 920°$

$200° = 200 - 360 = -160°$

$200° = 200 - 720 = -520°$

So far we talked about positive and negative angles, as well as angles beyond 360°. What about an angle that is a fraction of 1°?

Imagine if you move the terminal side half way between 0° and 1°; clearly, in decimal form, your angle is 0.5°. However, mathematicians further break down one degree into the form of minutes and seconds.

Just as in the clock, 1 hour = 60 minutes and 1 minute = 60 seconds, in Trigonometry, 1 degree = 60 minutes and 1 minute = 60 seconds.

Example 4:   Convert each angle measure to decimal form.
  a) 47° 52'   b) 100° 15' 35"   c) -65° 18"

Solution: You can use your calculator to convert angles in degrees, minutes, seconds into decimal form and vice versa. Look for the key that say D°M'S". Here is another way to do it.

a) $52' = 52' \times \dfrac{1°}{60'} = \dfrac{52}{60} = 0.87.$ So, $47° \ 52' = 47 + 0.87 = 47.87°$

b) $15' = 15' \times \dfrac{1°}{60'} = \dfrac{15}{60} = 0.25,$ and

$35" = 35" \times \dfrac{1'}{60''} \times \dfrac{1°}{60'} = \dfrac{35}{3600} = 0.0097.$

So, $100° \ 15' \ 35" = 100 + 0.25 + 0.0097 = 100.2597°$

c) $18" = \dfrac{18}{3600} = 0.005.$ So, $-65° \ 18" = -65.005°$

Example 5:   Convert each angle measure to D°M'S" form.
a) 71.4°   b) 10.23°   c) -1.624°

Solution: a) $0.4° = 0.4° \times \frac{60'}{1°} = 0.4 \times 60 = 24'$. So, $71.4° =$ $\boxed{71° \ 24'}$

b) $0.23° = 0.23° \times \frac{60'}{1°} = 0.23 \times 60 = 13.8' = 13' + 0.8'$, and

$0.8' = 0.8' \times \frac{60''}{1'} = 0.8 \times 60 = 48''$.

So, $10.23° =$ $\boxed{10° \ 13' \ 48''}$

c) $0.624° = 0.624° \times \frac{60'}{1°} = 0.624 \times 60 = 37.44' = 37' + 0.44'$, and

$0.44' = 0.44' \times \frac{60''}{1'} = 0.44 \times 60 = 26.4''$.

So, $-1.624° =$ $\boxed{-1° \ 37' \ 26''}$

Ok, now that you learned everything about angle measurement, let's revisit the arc length formula from Geometry. Recall that if you want to find the length of an arc on the circle, you take a fraction of the circumference.

For example, to find the arc length s for the figure at the right,

first find the circumference; $C = 2\pi r = 2\pi \times 10 = 20\pi$.

$s = \frac{60}{360} \times 20\pi = \frac{1}{6} \times 20\pi = \frac{10}{3}\pi \approx$ $\boxed{10.5}$

So in degree, any arc length $\boxed{s = \frac{\theta}{360°} \times 2\pi r}$

Since $360° = 2\pi$ in radian, this means $\boxed{s = \frac{\theta}{2\pi} \times 2\pi r = \theta \times r}$

For the above example, $\theta = 60° = \frac{\pi}{3}$. So, $s = \theta \times r = \frac{\pi}{3} \times 10 =$ $\boxed{10.5}$

Similarly, to find the area of a sector on the circle, you take a fraction of πr^2.

For example, to find the area of one slice of a pizza with an 8-inch radius, where each slice has a 45° central angle, you use

$$A = \frac{45°}{360°} \times \pi \times 8^2 = \frac{1}{8} \times 64 \times \pi = 8\pi = 25.12 \ in^2.$$

So in degree, $\boxed{A = \frac{\theta}{360°} \times \pi r^2.}$ In radian, $\boxed{A = \frac{\theta}{2\pi} \times \pi r^2 = \frac{\theta}{2} \times r^2}$

Ok I want to switch gear and finish this lesson with the concept of linear and angular speed. A lot of students are often confused about what they are.

You have probably learned in middle school that, distance = rate × time, or d = rt. For instance, if your parent drove, on average, at 50 miles per hour for 2 hours, then their distance is 100 miles.

Since d = rt, it means $r = \frac{d}{t}$. So, your rate or speed is distance traveled divided by time. Linear speed in Trigonometry is the same. If it takes you a certain amount of time t to rotate an arc length s, then the **linear speed** v is: $v = \frac{s}{t}$.

$$\text{Since s} = \theta \times r, \ \boxed{v = \frac{\theta r}{t}}$$

Similarly, if you take a certain amount of time t to sweep out an angle θ, then the **angular speed** is $\boxed{w = \frac{\theta}{t}}$

Finally, since $w = \frac{\theta}{t}$ and $v = \frac{\theta r}{t}$, we can relate the two and say $\boxed{v = w \times r}$

Example 6:   Suppose you are sitting on a ferris wheel that has a radius of 10 meter. It takes 4 minutes for the wheel to rotate 3.5 revolutions. What is the angular and linear speed of the wheel?

Solution: 1 revolution = 360° = 2 π radian.
                3.5 revolution = 3.5 × 2 π = 7 π radian.

$$w = \frac{\theta}{t} = \frac{7\pi \ rad}{4 \ min} = \boxed{5.5 \ rad/min}$$

$$v = wr = 5.5 \times 10 = \boxed{55 \ m/min.}$$

Practice 2

Convert the following from degrees to radians or vice versa.

1) 82°

2) 405°

3) -240°

4) $\dfrac{\pi}{8}$

5) $\dfrac{5\pi}{12}$

6) $\dfrac{17\pi}{6}$

Find two positive and two negative angles that are coterminal to each given angle.

7) 160°

8) -100°

9) 51.4°

10) $\dfrac{7\pi}{6}$

11) $\dfrac{-11\pi}{3}$

12) 1.5 rad

Convert each angle measure to decimal form.

13) 96° 21'

14) 53° 57' 6"

15) 30' 10"

Convert each angle measure to D°M'S" form.

16) 89.15°

17) -0.875°

18) 120.07°

19) Find the arc length s and area A.

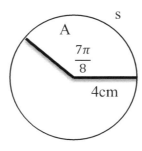

20) You are running on a circular track with a diameter of 70m. It takes you 100 second to finish 3 laps. What is the angular speed in radian per second? What is the linear speed in meter per second?

Did you know?

To multiply 2 two-digit numbers, where both numbers have the same first digit and their second digits add to 10, simply multiply the first digit by one plus itself and attach the product of their second digits at the end.

For example, 24 x 26 = **62**4 because 2 x 3 = 6 and 4 x 6 = 24.

Here is another example. What is 93 x 97 equal?

The answer is **90**21 because 9 x 10 = 90 and 3 x 7 = 21.

Lesson 3: The Unit Circle

Now that you are familiar with the radian angle measurement, I can talk more about the six Trigonometric functions.

Let's begin by reviewing special triangles from Geometry.

Recall that there are two types of special triangles.

1) $45° - 45° - 90°$ triangles:

Consider a square of length x.
If you draw a diagonal to cut the square in half, then the triangles are $45° - 45° - 90°$ triangles.

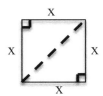

To find the hypotenuse c, you can use the Pythagorean Theorem.

$$c^2 = x^2 + x^2 \implies c^2 = 2x^2$$
$$c = \sqrt{2x^2} = \boxed{x\sqrt{2}}$$

So, the ratio in a $45° - 45° - 90°$ triangle is **x to x to $x\sqrt{2}$**.

Example 1: Find the missing length.

a)

b)

c)

Solution:
a) x = 7 cm, y = $7\sqrt{2}$ cm.

b) To go from the hypotenuse back to the shorter sides, you divide by $\sqrt{2}$. So, x = y = $5\sqrt{2} \div \sqrt{2}$ = 5.

c) x = y = $\frac{1}{\sqrt{2}} = \frac{1}{\sqrt{2}} \times \frac{\sqrt{2}}{\sqrt{2}} = \frac{\sqrt{2}}{2}$.

2) 30° - 60° - 90° triangles:

Consider an equilateral triangle of length 2x. If you "cut" the equilateral triangle in half, you get a 30° - 60° - 90° triangle.

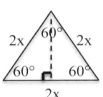

To find the height b, you can use the Pythagorean Theorem.

$$x^2 + b^2 = (2x)^2 \Rightarrow x^2 + b^2 = 4x^2$$
$$\Rightarrow b^2 = 3x^2$$
$$\Rightarrow b = \sqrt{3x^2}$$
$$\Rightarrow b = \boxed{x\sqrt{3}}$$

So, the ratio in a 30° - 60° - 90° triangle is **x to $x\sqrt{3}$ to 2x**.

Example 2: Find the missing length.

a)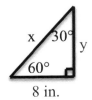

b)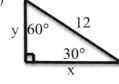

c)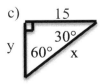

19

Solution:

a) Since the side opposite the 30 degree angle is given, this problem is easy. $x = 2 \times 8 = 16$ in. $y = 8\sqrt{3}$ in.

b) Since the side opposite the 30 degree angle is NOT given, you need to find it first. It is half the hypotenuse.
So, $y = \frac{1}{2} \times 12 = 6$, and $x = 6\sqrt{3}$.

c) Again we need to find the side opposite 30 degree angle first.
$15 = y\sqrt{3} \Rightarrow y = \frac{15}{\sqrt{3}} = \frac{15\sqrt{3}}{3} = 5\sqrt{3}$, and $x = 2y = 2 \times 5\sqrt{3} = 10\sqrt{3}$.

Now let's apply what we learned about special triangles to the "unit" circle, which is a circle of radius "one."

Given the unit circle shown at the right, if you rotate the terminal side $30°$ or $\frac{\pi}{6}$, what is x and y?

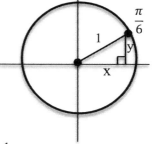

Since y is opposite $30°$, $\boxed{y = \frac{1}{2}}$

$x = \frac{1}{2} \times \sqrt{3} = \boxed{\frac{\sqrt{3}}{2}}$

Here is how to fill out the first quadrant of the unit circle.

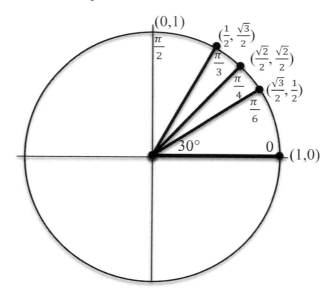

Now if you rotate the terminal side 150°, it will make a 30° angle with the x-axis in quadrant II. So, instead of $(\frac{\sqrt{3}}{2}, \frac{1}{2})$, the point on the circle will be $(-\frac{\sqrt{3}}{2}, \frac{1}{2})$.

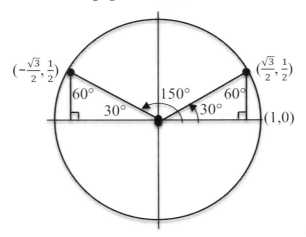

What about quadrant III? Well, both the x- and y-values will be negative, because the points are on the left and below the x-axis.

In quadrant IV, x-values will be positive; y-values will be negative, because the points are on the right but below the x-axis.

Here is the whole unit circle.

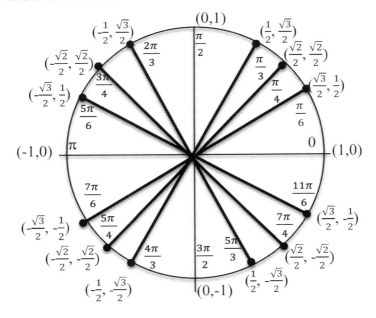

You do not need to memorize the whole unit circle. Just remember the correct order for 30° is $(\frac{\sqrt{3}}{2}, \frac{1}{2})$, then for 60° you reverse the order. For 45°, it is simple, because the x- and y-coordinates are the same. For the other three quadrants, you just change the signs of x- or y-coordinates accordingly. If you are on the left, namely quadrant II or III, x-values are negative. If you are below the x-axis, namely quadrant III or IV, y-values are negative.

Here is the sign chart if you want to visualize it.

$$(-,+) \qquad (+,+)$$
$$(-,-) \qquad (+,-)$$

Important: What about when the angles are not special?

If I move the terminal side to a random angle, for example θ = 50°, what are the x- and y-coordinates?

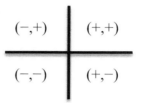

Since
$$\sin\theta = \frac{opposite}{hypotenuse},$$

$$\sin\theta = \frac{y}{1} = y.$$

Similarly,
$$\cos\theta = \frac{x}{1} = x.$$

This means, to find the x-value, you just take cosine of the angle, whether it is special or not. To find the y-value, you take sine. Conversely, to find sine of an angle, you look at its y-value.

So from the unit circle, $\sin 30° = \sin\frac{\pi}{6} = \frac{1}{2}$; $\cos 30° = \cos\frac{\pi}{6} = \frac{\sqrt{3}}{2}$.

Although you only need to remember 30° and 45° angles to fill out the whole unit circle, I came up with an even better trick to memorize sine, cosine, and tangent of special angles.

But first notice that if you take sin ÷ cos, you will get tan, because

$$\frac{sin\theta}{cos\theta} = \frac{opposite}{hypotenuse} \div \frac{adjacent}{hypotenuse} = \frac{opposite}{hypotenuse} \times \frac{hypotenuse}{adjacent} = \frac{opposite}{adjacent} = tan\theta.$$

Here is the trick.

1) List the 5 angles in quadrant I in order horizontally. (See the table below)

2) Write down "sin, cos, tan" vertically on the left. (Hint: sin over cos is tan)

3) Fill out the first row starting with $\frac{\sqrt{0}}{2}$ and count up to $\frac{\sqrt{4}}{2}$.

4) Reverse the first row for the second row.

5) Divide the first row by the second row to get the third row.

	0°	30°	45°	60°	90°
sin	$\frac{\sqrt{0}}{2}$	$\frac{\sqrt{1}}{2}$	$\frac{\sqrt{2}}{2}$	$\frac{\sqrt{3}}{2}$	$\frac{\sqrt{4}}{2}$
cos	$\frac{\sqrt{4}}{2}$	$\frac{\sqrt{3}}{2}$	$\frac{\sqrt{2}}{2}$	$\frac{\sqrt{1}}{2}$	$\frac{\sqrt{0}}{2}$
tan	0	$\frac{1}{\sqrt{3}}$	1	$\sqrt{3}$	undefined

With a little bit of practice, it should take you less than a minute to re-create the above table. You can then use it to look up values for your entire tests.

Warning: Wait until after your teacher passed out the test, then create the table on scratch paper. Do not bring it into a test, that would be consider cheating.

The benefit of using this table instead of the unit circle is, it gives you the values of tangent as well. But to fully take advantage of this trick, you need to know the concept of **reference angles**. These are acute angles that the terminal side makes with the x-axis. So for example, to find sin300°, you use 60° as your "reference." From the table, $\sin 60° = \frac{\sqrt{3}}{2}$, and since 300° is in quadrant IV, $\sin 300° = -\frac{\sqrt{3}}{2}$.

Basically, just use the table first and then determine whether you need to change the sign of the value or not. You can use the sign chart we have earlier, or you can try to remember the acronym: **A**ll **S**tudents **T**ake **C**alculus.

Students	**A**ll	
Take	**C**alculus	

meaning

Sin is positive	All positive
Tan is positive	Cos is positive

Example 3: Find the exact value of each of the following.

a) $\sin 225°$ b) $\cos\dfrac{11\pi}{6}$ c) $\sin\dfrac{5\pi}{6}$ d) $\tan\dfrac{3\pi}{4}$

Solution: a) $225°$ is $45°$ in quadrant III; so sin is negative $\Rightarrow \sin 225° = -\dfrac{\sqrt{2}}{2}$.

b) $\dfrac{11\pi}{6}$ is $\dfrac{\pi}{6}$ in quadrant IV; so cos is positive $\Rightarrow \cos\dfrac{11\pi}{6} = \dfrac{\sqrt{3}}{2}$.

c) $\dfrac{5\pi}{6}$ is $\dfrac{\pi}{6}$ in quadrant II; so sin is positive $\Rightarrow \sin\dfrac{5\pi}{6} = \dfrac{1}{2}$.

d) $\dfrac{3\pi}{4}$ is $\dfrac{\pi}{4}$ in quadrant II; so tan is negative. This means $\tan\dfrac{3\pi}{4} = -1$.

That's it. Make sure you practice. From my experience, Trigonometry can quickly overwhelm students. You need to master this lesson before moving on.

Practice 3

1) Create and fill out the entire unit circle.

2) Create and fill out "the table." (from memory of course).

3) Find the length of the diagonal of a square with sides of 24 cm.

4) Find the length of the height of an equilateral triangle with sides of 11 ft.

Find the exact value of each of the following.

5) $\sin \frac{7\pi}{6}$ 6) $\cos 510°$

7) $\tan \frac{\pi}{2}$ 8) $\cos \pi$

9) $\sin \frac{3\pi}{2}$ 10) $\tan \pi$

11) $\csc \frac{3\pi}{4}$ 12) $\sec \frac{11\pi}{6}$

13) $\sec \pi$ 14) $\cot \frac{5\pi}{4}$

15) $\cot \frac{4\pi}{3}$ 16) $\csc -390°$

17) $\tan \frac{-5\pi}{6}$ 18) $\sec \frac{8\pi}{3}$

19) $\csc 0$ 20) $\cot -\pi$

Bonus: Find the exact value.

$$\sin \frac{\pi}{6} + \cos \frac{-2\pi}{6} + \cot \frac{3\pi}{6} + \tan \frac{-4\pi}{6} + \csc \frac{5\pi}{6} + \sec \frac{-6\pi}{6}$$

Did you know?

If you have a right triangle with the hypotenuse and one of the legs given, you can take advantage of the difference of squares formula, which is $x^2 - y^2 = (x - y)(x + y)$, to help you quickly find the missing leg.

For example, suppose the hypotenuse $c = 36$, and one of the legs $b = 26$.

According to the Pythagorean Theorem, $a^2 + b^2 = c^2$.

So, $a^2 = c^2 - b^2 = (c - b)(c + b) = (36 - 26)(36 + 26) = 10 \times 62 = 620$.

This means $a = \sqrt{620}$. This is faster than calculating $36^2 - 26^2$ directly.

Lesson 4: Trigonometric Graphs, Part One

Let's begin with an example of the simplest Trigonometric graph.

Example 1: Graph $y = \sin x$.

Solution: When you are graphing a new function that you have never seen before, you can always start by picking points.

What is y when x equals 0 radian? What about when x is $\frac{\pi}{2}$ rad?

Here is the table of values.

X	0	$\frac{\pi}{2}$	π	$\frac{3\pi}{2}$	2π
Y	0	1	0	-1	0

Ok let's plot these points.

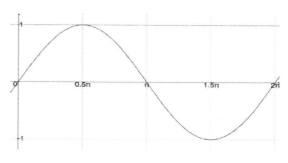

You can see that the sine curve starts at 0, increases to 1, decrease to 0, further decrease to -1, and increases back to 0. It is very important that you remember this "sine curve" pattern, because later on it will allow you to quickly sketch sine graphs without picking points.

Now, what if you pick more points to the left or right of the graph? Well, the graph just "repeats" itself, because it is **periodic**. In fact, we have just graphed one period of sinx. Let's graph another period on the left. In other words, let's pick negative angles this time.

Here is the table of values.

X	$\frac{-\pi}{2}$	$-\pi$	$\frac{-3\pi}{2}$	-2π
Y	-1	0	1	0

Ok let's add these additional points to the previous graph.

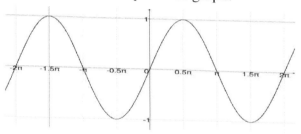

Example 2:   Graph $y = \sin(x + \frac{\pi}{4})$.

Solution: In the previous example, we started out by picking 0 radian. This time we will pick $x = \frac{-\pi}{4}$, because that is equivalent to picking $x = 0$ last time. That is, $y = \sin(x + \frac{\pi}{4}) = \sin(\frac{-\pi}{4} + \frac{\pi}{4}) = \sin(0) = 0$.

Since sine has a period of 2π, if we start at $\frac{-\pi}{4}$, we will end at $\frac{-\pi}{4} + 2\pi = \frac{7\pi}{4}$.

To find the middle point, take the average of $\frac{-\pi}{4}$ and $\frac{7\pi}{4}$; so we have $(\frac{-\pi}{4} + \frac{7\pi}{4}) \div 2 = \frac{3\pi}{4}$.

Taking the average of $\frac{-\pi}{4}$ and $\frac{3\pi}{4}$, as well as $\frac{3\pi}{4}$ and $\frac{7\pi}{4}$ give the last two points.

Here is the table of values.

X	$\dfrac{-\pi}{4}$	$\dfrac{\pi}{4}$	$\dfrac{3\pi}{4}$	$\dfrac{5\pi}{4}$	$\dfrac{7\pi}{4}$
Y	0	1	0	-1	0

Here is the graph.

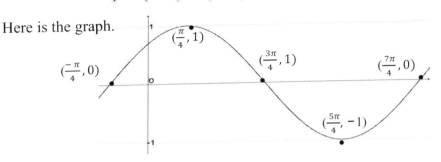

Notice that the graph of $y = \sin(x + \frac{\pi}{4})$ is the same as the graph of $y = \sin x$ shifted to the left by $\frac{\pi}{4}$.

Now, is every sin graph has a period of 2π? The answer is "No."

Important: In general, the graph of $y = a\sin(bx + c) + d$ has a period of $\dfrac{2\pi}{b}$. Also, to find the **phase shift**, set $bx + c = 0$ and solve for x.

Example 3:   Graph $y = 2\sin(\frac{1}{2}x - \frac{\pi}{6})$.

Solution: First find the period using the formula $\dfrac{2\pi}{b}$.

Since $b = \frac{1}{2}$, we have $2\pi \div \frac{1}{2} = 2\pi \times \frac{2}{1} = 4\pi$.

Next find the phase shift. $\frac{1}{2}x - \frac{\pi}{6} = 0 \Rightarrow x = \frac{\pi}{3}$.

So, the starting point is $\frac{\pi}{3}$; the end point is $\frac{\pi}{3} + 4\pi = \frac{13\pi}{3}$.

The middle is $(\frac{\pi}{3} + \frac{13\pi}{3}) \div 2 = \frac{7\pi}{3}$.

The other 2 points are: $(\frac{\pi}{3} + \frac{7\pi}{3}) \div 2 = \frac{4\pi}{3}$ and $(\frac{7\pi}{3} + \frac{13\pi}{3}) \div 2 = \frac{10\pi}{3}$.

Here is the table of values.

X	$\frac{\pi}{3}$	$\frac{4\pi}{3}$	$\frac{7\pi}{3}$	$\frac{10\pi}{3}$	$\frac{13\pi}{3}$
Y	0	2	0	-2	0

Here is the graph.

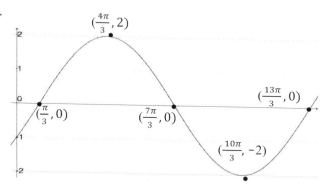

Notice every sine graphs have the same shape, except you just label the x-axis differently. Also, the highest and lowest points on this graph are 2 and -2, not 1 and -1. In fact, half the difference between the highest and lowest points is called the **amplitude**. In this example, the amplitude is $(2 - {-2}) \div 2 = 2$.

Example 4:   Graph two periods of $y = \cos x$.

Solution: Again, if we start with $x = 0$ radian, then $y = \cos(0) = 1$.

Here is the table of values.

X	0	$\frac{\pi}{2}$	π	$\frac{3\pi}{2}$	2π	$\frac{5\pi}{2}$	3π	$\frac{7\pi}{2}$	4π
Y	1	0	-1	0	1	0	-1	0	1

Here is the graph.

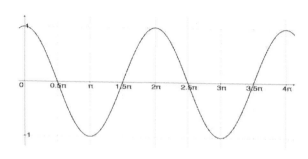

Note that instead of graphing two periods from -2π to 2π, I decided to graph from 0 to 4π to show you a second way that other people might do it. Also, comparing to the sine curve, one period of cosine starts at 1, decreases to 0, further decrease to -1, increase to 0, and further increases back to 1.

Example 5: Graph two periods of y = -cosx + 2.

Solution: Start with x = 0, then y = -cos(0) + 2 = -1 + 2 = 1.

Here is the rest of the points in the table.

X	0	$\frac{\pi}{2}$	π	$\frac{3\pi}{2}$	2π	$\frac{-\pi}{2}$	$-\pi$	$\frac{-3\pi}{2}$	-2π
Y	1	2	3	2	1	2	3	2	1

Here is the graph.

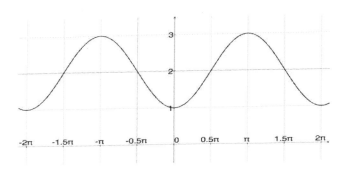

Note that the negative in front of cosx makes the graph reflect upside down; the "2" shifts the graph up 2 units. The amplitude is (3 - 1) ÷ 2 = 1.

Again, in general if you have y = acos(bx + c) + d, then the period is $\frac{2\pi}{b}$; the phase shift is bx + c = 0 or x = -c/b.

Example 6:   Graph $y = 2\cos(\frac{\pi}{4}x + \frac{3\pi}{4}) + 1$.

Solution: First, the period is $\frac{2\pi}{b} = 2\pi \div \frac{\pi}{4} = \frac{2\pi}{1} \times \frac{4}{\pi} = 8$.

Second, the phase shift is $\frac{\pi}{4}x + \frac{3\pi}{4} = 0 \;\Rightarrow\; \frac{\pi}{4}x = \frac{-3\pi}{4}$

$$x = \frac{-3\pi}{4} \times \frac{4}{\pi} = -3.$$

So, one period starts at x = -3 and end at x = -3 + 8 = 5.

Also a = 2; so the amplitude is 2.

Finally, d = 1; so the vertical translation is up 1 unit.

With some practices, you should be able to skip the table of values and graph sine or cosine according to the values of a, b, c, and d.

Remember: cosine starts at the highest y-value, decreases to the middle, further decreases to the lowest point, increases to the middle, further increases back to the top.

Here is the graph.

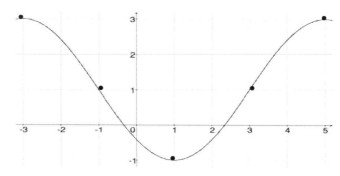

Notice that the x-axis does not contain "π," since the angle $(\frac{\pi}{4}x + \frac{3\pi}{4})$ already has "π" in it.

Ok, that's it. It's your turn. Have fun graphing. See you next lesson.

Practice 4

Without graphing, find the amplitude, period, phase shift and vertical translation.

1) $y = \sin(x + \frac{\pi}{8})$

2) $y = -\cos(2x - \frac{3\pi}{4})$

3) $y = 2\cos(\pi x) + 1$

4) $y = \frac{9}{5}\sin(\frac{7\pi}{12}x + \frac{\pi}{10}) - 3$

5) $y = 2 - \sin(2x + \frac{\pi}{3})$

6) $y = 3 + \frac{3}{4}\cos(\frac{\pi}{9}x - \pi)$

Graph two periods for each of the following.

7) $y = \sin(x + \frac{\pi}{3})$

8) $y = -\sin(x) - 2$

9) $y = \frac{1}{2}\cos(2x)$

10) $y = 1 + 2\sin(3x)$

11) $y = -\cos(\frac{1}{4}x - \frac{\pi}{6})$

12) $y = 3\cos(\frac{5\pi}{3}x) - 3$

13) $y = \sin(\frac{2}{3}x - \frac{4\pi}{3})$

14) $y = 2 - \cos(x - \frac{7\pi}{4})$

15) $y = \sin(\pi x + \frac{\pi}{2}) + 3$

16) $y = 4\sin(\frac{2\pi}{5}x + \pi) + 2$

17) $y = \cos(x - \frac{3\pi}{2})$

18) $y = -2\sin(4x - \frac{2\pi}{3}) - 1$

19) Given $y = 2\cos(x) + 2$ and $-2\pi \le x \le 2\pi$, find the x- and y-intercept.

20) Given $y = 3\sin(x) - 5$ and $0 \le x \le 4\pi$, how many time does the graph intersect the line $y = -8$?

Bonus: Graph two periods for each of the following.

a) $y = \sin x + \cos x$ b) $y = \cos x - \sin x$ c) $y = 4\sin x \cos x$

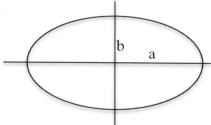

Did you know?

To find the area of a circle with radius **r**, *you use the formula* **A** = π**r²** = π x **r** x **r**.
How about finding the area of an ellipse
or an oval shown at the right?
The formula is very similar.
It is **A** = π x **a** x **b**.

Lesson 5: Trigonometric Graphs, Part Two

In this lesson, we will continue to graph the remaining Trigonometric functions.
We will begin with y = tanx.

Example 1: Graph two periods of y = tan(x).

Solution: Recall that $tan(x) = \frac{sin(x)}{cos(x)}$. So, whenever cos(x) equals 0, tan(x) is
undefined. Cos(x) = 0 at x equals . . . $\frac{-\pi}{2}$, $\frac{\pi}{2}$, $\frac{3\pi}{2}$, . . .
Therefore, tan(x) is undefined at odd multiples of $\frac{\pi}{2}$.

Next, tan(x) = 0 whenever sin(x) = 0.
sin(x) = 0 at x equals . . . - π, 0, π, 2π . . .
Therefore, tan(x) is zero whenever x = $n\pi$, for any integer n.

Finally, tan($\frac{\pi}{4}$) = 1, and tan($\frac{-\pi}{4}$) = -1.

Here is the table of values for one period of tan(x).

X	$\frac{-\pi}{2}$	$\frac{-\pi}{4}$	0	$\frac{\pi}{4}$	$\frac{\pi}{2}$
Y	undefined	-1	0	1	undefined

Here is the graph for two periods of tan(x).

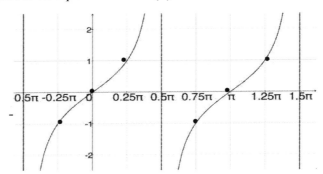

Notice that tan(x) "repeats itself" after 1π, NOT 2π like sin(x) and cos(x).

Important: In general, the graph of y = atan(bx + c) + d, or y = acot(bx + c) + d has a period of $\frac{\pi}{b}$.

Example 2: Graph two periods of y = cot(x).

Solution: Cot(x) is the reciprocal of tan(x). So, $\cot(x) = \frac{\cos(x)}{\sin(x)}$.

This means cot(x) is undefined whenever sin(x) = 0, namely x = . . . - π, 0, π, 2π, . . .

Cot(x) = 0 whenever cos(x) = 0, namely, x = . . . $\frac{-\pi}{2}$, $\frac{\pi}{2}$, $\frac{3\pi}{2}$, . . .

Here is the table of values for one period of cot(x).

X	0	$\frac{\pi}{4}$	$\frac{\pi}{2}$	$\frac{3\pi}{4}$	π
Y	undefined	1	0	-1	undefined

Here is the graph for two periods of cot(x).

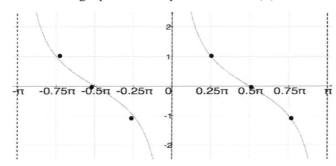

Example 3:   Graph $y = \tan(\frac{1}{2}x - \frac{\pi}{6})$.

Solution: When graphing sin and cos, the most important things are the periods and the phase shifts. When graphing tan and cot, it's more important to first figure out where they are undefined.

Since $\tan(\frac{\pi}{2})$ is undefined, we can find the first undefined or "vertical asymptote" by setting the angle $\frac{1}{2}x - \frac{\pi}{6} = \frac{\pi}{2}$ and solve for x.

$$\frac{1}{2}x - \frac{\pi}{6} = \frac{\pi}{2} \implies \frac{1}{2}x = \frac{\pi}{2} + \frac{\pi}{6} = \frac{4\pi}{6}$$

$$\implies x = 2 \times \frac{4\pi}{6} = \frac{4\pi}{3}.$$

Also, the period is $\frac{\pi}{b} = \frac{\pi}{1/2} = 2\pi$.

The second vertical asymptote is $\frac{4\pi}{3} + 2\pi = \frac{10\pi}{3}$.

The third vertical asymptote is $\frac{4\pi}{3} - 2\pi = \frac{-2\pi}{3}$.

The x-intercepts are the middles of the asymptotes, which are

$$\frac{\frac{4\pi}{3} + \frac{10\pi}{3}}{2} = \frac{14\pi}{3} \times \frac{1}{2} = \frac{7\pi}{3}, \text{ and } \frac{\frac{4\pi}{3} + \frac{-2\pi}{3}}{2} = \frac{2\pi}{3} \times \frac{1}{2} = \frac{\pi}{3}.$$

So far we have the following.

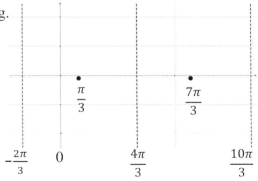

To find the rest of the points, average the middles and the asymptotes. For example, $x = \frac{\frac{\pi}{3} + \frac{4\pi}{3}}{2} = \frac{5\pi}{3} \times \frac{1}{2} = \frac{5\pi}{6}$.

$y = \tan(\frac{1}{2}x - \frac{\pi}{6}) = \tan(\frac{1}{2} \times \frac{5\pi}{6} - \frac{\pi}{6}) = \tan(\frac{3\pi}{12}) = \tan(\frac{\pi}{4}) = 1$.

Here is the whole graph.

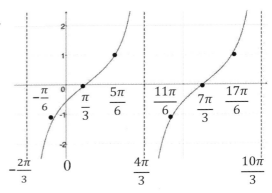

Example 4: Graph $y = 2\cot(\frac{\pi}{2}x - \frac{\pi}{4})$.

Solution: Since $\cot(0) = \frac{cos(0)}{sin(0)} = \frac{1}{0}$ is undefined, set $\frac{\pi}{2}x - \frac{\pi}{4} = 0$.

$$\Rightarrow \quad \frac{\pi}{2}x = \frac{\pi}{4}$$
$$\Rightarrow \quad x = \frac{1}{2}.$$

Also, the period is $\frac{\pi}{b} = \frac{\pi}{\pi/2} = \pi \times \frac{2}{\pi} = 2$.

So the vertical asymptotes are: $x = \frac{1}{2}$, $x = \frac{1}{2} + 2 = \frac{5}{2}$, $x = \frac{1}{2} - 2 = \frac{-3}{2}$.

The x-intercepts are: $x = (\frac{-3}{2} + \frac{1}{2}) \div 2 = \frac{-1}{2}$, and $x = (\frac{1}{2} + \frac{5}{2}) \div 2 = \frac{3}{2}$.

Here is the whole graph.

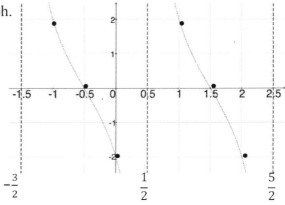

Ok, let's finish this lesson with the graphs of csc(x) and sec(x).

Example 5: Graph $y = \csc(x)$.

Solution: $\csc(x) = \frac{1}{\sin(x)}$. For example, when $x = \frac{\pi}{6}$, $\csc(\frac{\pi}{6}) = \frac{1}{\sin(\frac{\pi}{6})} = \frac{1}{1/2} = 2$.

Here is the table of values for csc(x), along with sin(x).

X	0	$\frac{\pi}{6}$	$\frac{\pi}{2}$	$\frac{5\pi}{6}$	π	$\frac{7\pi}{6}$	$\frac{3\pi}{2}$	$\frac{11\pi}{6}$
sin(x)	0	$\frac{1}{2}$	1	$\frac{1}{2}$	0	$\frac{-1}{2}$	-1	$\frac{-1}{2}$
csc(x)	undefined	2	1	2	undefined	-2	-1	-2

Here is the graph of csc(x), along with sin(x) in dotted line.

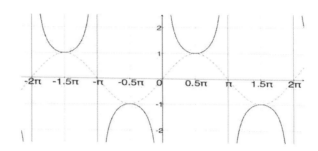

Notice that csc(x) "repeats itself" every 2π. Also, the lower the values of sin(x), the higher the values of csc(x).

Example 6: Graph $y = \sec(x)$.

Solution: $\sec(x) = \frac{1}{\cos(x)}$. So, using the same technique as in example 5, you can just graph cos(x) first, and then reflects it to get sec(x).

Here is the graph of sec(x), along with cos(x) in dotted line.

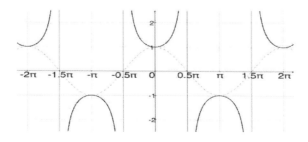

Example 7:   Graph $y = \csc(2x - \frac{\pi}{2})$.

Solution: First graph $\sin(2x - \frac{\pi}{2})$. The period is $\frac{2\pi}{2} = \pi$. The phase shift is

$$2x - \frac{\pi}{2} = 0 \implies x = \frac{\pi}{4}.$$

One period starts at $x = \frac{\pi}{4}$ and end at $x = \frac{\pi}{4} + \pi = \frac{5\pi}{4}$.

Here is the graph of two periods of $y = \csc(2x - \frac{\pi}{2})$, along with $y = \sin(2x - \frac{\pi}{2})$ in dotted line.

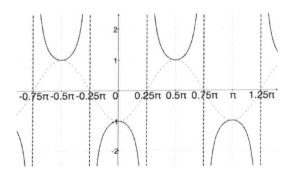

Ok it's your turn. Enjoy the practice. See you next lesson.

Practice 5

List all x-values, where $-2\pi \le x \le 2\pi$, for which the given function is undefined.

1) $y = \tan(x - \frac{\pi}{4})$

2) $y = 5\cot(x - \frac{\pi}{3})$

3) $y = -\tan(2x + \frac{\pi}{6}) + 1$

4) $y = \csc(x + \frac{\pi}{4}) - 2$

5) $y = \frac{3}{5}\sec(3x - \frac{\pi}{2}) - 4$

6) $y = \cot(\frac{2}{9}x + \frac{5\pi}{6}) - 1$

7) Given $y = -\frac{1}{6}\csc(x - \frac{5\pi}{4})$, what is the y-intercept?

8) Given $y = 2\cot(\frac{\pi}{8}x - \frac{1}{2})$, what is the period?

Graph each of the following.

9) $y = \tan(x + \frac{\pi}{3})$

10) $y = -\tan(x)$

11) $y = -\cot(\frac{1}{2}x)$

12) $y = \cot(\frac{\pi}{3}x - \frac{\pi}{3})$

13) $y = \csc(3x)$

14) $y = \sec(\frac{1}{3}x + \pi)$

15) $y = \frac{1}{2}\tan(2x - \frac{\pi}{6})$

16) $y = -\csc(x)$

17) $y = -\sec(x) - 2$

18) $y = 2\csc(x) - 1$

19) $y = -3\cot(\frac{2}{3}x + 4) + 1$

20) $y = 2 - \sec(\frac{\pi}{4}x - \frac{\pi}{3})$

Bonus: Graph two periods for each of the following.

a) $y = |tanx| + 1$　　　b) $y = |2 - secx|$　　　c) $y = -|cscx| + 1$

> ### *Did you know?*
>
>   *To see if a number is divisible by 3, you can simply add up all its digits. If the sum is divisible by 3, then the original number is divisible by 3.*
>   *For example, 24156 is divisible by 3 because 2 + 4 + 1 + 5 + 6 = 18, and 18 is divisible by 3.*
>   *Here is another example. 932510 is **not** divisible by 3 because 9 + 3 + 2 + 5 + 1 + 0 = 20, and 20 is not divisible by 3.*

Lesson 6: Inverse Trigonometric Functions

Recall from Algebra that a relation is a function if for each x-value there is at most one y-value, or equivalently, the graph passes the vertical line test. Furthermore, for a function to have an inverse, it must be "one-to-one." That is, for each x-value there is one y-value, and for each y-value there is one x-value. In other words, the graph must pass BOTH the vertical and horizontal line tests.

From the previous lesson, you can see that the graph of $y = \sin(x)$ does not pass the horizontal line test. So, it is not invertible.

For example, even though $\sin 30° = \frac{1}{2}$, you **cannot** automatically assume "sine inverse of $\frac{1}{2}$," written as $\sin^{-1}(\frac{1}{2})$ *or* $arcsin(\frac{1}{2})$, equals $30°$ or $\frac{\pi}{6}$.

If you look further, you will see that $\sin 150°$ also equals $\frac{1}{2}$; so is $\sin 390°$, and infinitely many other angles. How do we know $\sin^{-1}(\frac{1}{2}) \neq 150°$?

The answer is, mathematicians must decide and choose what the acceptable answer must be if they insist on studying inverse trigonometric functions.

As it turned out, they chose to restrict the domain of sine to the interval $[-\frac{\pi}{2}, \frac{\pi}{2}]$ so that the graph is one-to-one.

Here is the sine graph again.

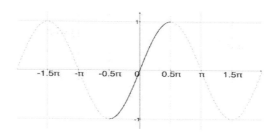

Here is the definition of sine inverse:

$$y = \sin^{-1}(x) \text{ if } x = \sin(y), \text{ where } -1 \le x \le 1 \text{ and } -\frac{\pi}{2} \le y \le \frac{\pi}{2}.$$

Based on this definition, $\sin^{-1}\left(\frac{1}{2}\right) = \frac{\pi}{6}$.

Example 1: Calculate: a) $\sin^{-1}\left(\frac{\sqrt{2}}{2}\right)$ b) $\sin^{-1}\left(\frac{-\sqrt{3}}{2}\right)$ c) $\sin^{-1}(0.8)$

Solution: a) $\sin^{-1}\left(\frac{\sqrt{2}}{2}\right) = \frac{\pi}{4}$, because $\sin\left(\frac{\pi}{4}\right) = \frac{\sqrt{2}}{2}$, and

$\frac{\pi}{4}$ is within the interval $[-\frac{\pi}{2}, \frac{\pi}{2}]$.

b) $\sin^{-1}\left(\frac{-\sqrt{3}}{2}\right) = \frac{-\pi}{3}$, because $\sin\left(\frac{-\pi}{3}\right) = \frac{-\sqrt{3}}{2}$, and

$\frac{-\pi}{3}$ is within the interval $[-\frac{\pi}{2}, \frac{\pi}{2}]$.

c) Since 0.8 is not one of the special values on the unit circle, you need to use a calculator for this one. You can put your calculator in degree or radian **mode**. You should get $\sin^{-1}(0.8) = 53.13°$ or 0.93 rad.

Warning: The most common mistake students make for part b is that they say the answer equals $\frac{5\pi}{3}$. Even though $\frac{5\pi}{3}$ and $\frac{-\pi}{3}$ are coterminal angles (i.e. both angles have the same terminal side in quadrant IV), $\frac{5\pi}{3}$ is NOT within the interval $[-\frac{\pi}{2}, \frac{\pi}{2}]$. So it is not an acceptable answer.

Next, let's look at cosine inverse. Here is the graph of cosine again.

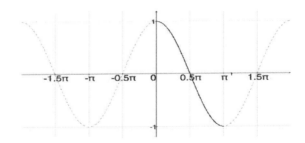

Since the cosine graph is the same as sine shifted 90° or $\frac{\pi}{2}$, it is natural to define inverse cosine as follow:

$$y = \cos^{-1}(x) \text{ if } x = \cos(y), \text{ where } -1 \le x \le 1 \text{ and } 0 \le y \le \pi.$$

Example 2:   Calculate: a) $\cos^{-1}(\frac{1}{2})$ b) $\cos^{-1}(\frac{-\sqrt{2}}{2})$ c) $\cos^{-1}(-\frac{1}{2})$

Solution: a) $\cos^{-1}(\frac{1}{2}) = \frac{\pi}{3}$, because $\cos(\frac{\pi}{3}) = \frac{1}{2}$ and $0 \le \frac{\pi}{3} \le \pi$.

b) $\cos^{-1}(\frac{-\sqrt{2}}{2}) = \frac{3\pi}{4}$, because $\cos(\frac{3\pi}{4}) = \frac{-\sqrt{2}}{2}$ and $0 \le \frac{3\pi}{4} \le \pi$.

c) $\cos^{-1}(-\frac{1}{2}) = \frac{2\pi}{3}$, because $\cos(\frac{2\pi}{3}) = -\frac{1}{2}$ and $0 \le \frac{2\pi}{3} \le \pi$.

Tips: For sine inverse, if the input is negative, you just need to put the negative sign in your answer. For example, $\sin^{-1}(\frac{1}{2}) = \frac{\pi}{6}$; $\sin^{-1}(-\frac{1}{2}) = \frac{-\pi}{6}$.

For cosine inverse, if the input is negative, your answer must be an angle in quadrant II. For example, $\cos^{-1}(\frac{1}{2}) = \frac{\pi}{3}$; $\cos^{-1}(-\frac{1}{2}) = \frac{2\pi}{3}$.

Here is a summary of how all six inverse trigonometric functions are defined:

1) $y = \sin^{-1}(x)$ if $x = \sin(y)$, where $-1 \le x \le 1$ and $-\frac{\pi}{2} \le y \le \frac{\pi}{2}$.

2) $y = \cos^{-1}(x)$ if $x = \cos(y)$, where $-1 \le x \le 1$ and $0 \le y \le \pi$.

3) $y = \tan^{-1}(x)$ if $x = \tan(y)$, where $-\infty < x < \infty$ and $-\frac{\pi}{2} < y < \frac{\pi}{2}$.

4) $y = \cot^{-1}(x)$ if $x = \cot(y)$, where $-\infty < x < \infty$ and $0 < y < \pi$.

5) $y = \sec^{-1}(x)$ if $x = \sec(y)$, where $|x| \ge 1$ and $0 \le y \le \pi$, $y \ne \frac{\pi}{2}$.

6) $y = \csc^{-1}(x)$ if $x = \csc(y)$, where $|x| \ge 1$ and $\frac{-\pi}{2} \le y \le \frac{\pi}{2}$, $y \ne 0$.

You can use the following figures to help you visualize the range of the first three inverse functions.

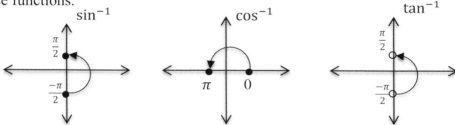

For the last three inverse functions, you should review their graphs from the previous lesson if you are not sure about the definitions.

Example 3:   Find each of the following.

   a) $\tan^{-1}(-1)$ b) $\sec^{-1}(2)$ c) $\csc^{-1}(-\sqrt{2})$ d) $\cos^{-1}(1.5)$

Solution: a) $\tan^{-1}(-1) = \frac{-\pi}{4}$, because $\tan(\frac{-\pi}{4}) = -1$ and $-\frac{\pi}{2} < \frac{-\pi}{4} < \frac{\pi}{2}$.

   b) Let $y = \sec^{-1}(2)$. This means $\sec(y) = 2$.

   This also means $\cos(y) = \frac{1}{2}$. So, $\boxed{y = \frac{\pi}{3}}$

43

c) Let $y = \csc^{-1}(-\sqrt{2})$. This means $\csc(y) = -\sqrt{2}$.

This also means $\sin(y) = \frac{-1}{\sqrt{2}}$. So, $\boxed{y = \frac{-\pi}{4}}$

d) If you put $\cos^{-1}(1.5)$ in your calculator, it should say **error**, because 1.5 is NOT in the domain of inverse cosine. The input must be within the interval [-1, 1].

Example 4: Graph $y = \tan^{-1}(x)$.

Solution: Here is the table of values for $\tan(x)$ from the previous lesson.

X	$\frac{-\pi}{2}$	$\frac{-\pi}{4}$	0	$\frac{\pi}{4}$	$\frac{\pi}{2}$
Y	undefined	-1	0	1	undefined

Since you are graphing inverse tan, just switch the x- and y-values.

Here is the table of values for $\tan^{-1}(x)$.

Y	$\frac{-\pi}{2}$	$\frac{-\pi}{4}$	0	$\frac{\pi}{4}$	$\frac{\pi}{2}$
X	undefined	-1	0	1	undefined

Also from the definition in this lesson, the domain for $\tan^{-1}(x)$ is $-\infty < x < \infty$. Here is the graph.

Note:
The vertical asymptote of $\tan(x)$ becomes the horizontal asymptote of $\tan^{-1}(x)$.

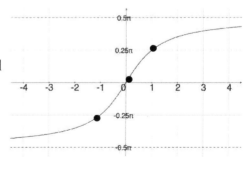

Ok, I will now talk about composite functions. Recall that when you compose a function and its inverse, you get the original input. That is, $f(f^{-1}(x)) = x$ and $f^{-1}(f(x)) = x$.

So, this means $\sin(\sin^{-1}(x)) = x$ if $-1 \le x \le 1$, and $\sin^{-1}(\sin(y)) = y$ if $-\frac{\pi}{2} \le y \le \frac{\pi}{2}$. The same idea applied for all other Trigonometric functions.

Example 5: Calculate each of the following.

a) $\cos(\cos^{-1}(\frac{1}{2}))$ b) $\tan^{-1}(\tan(\frac{\pi}{6}))$ c) $\cot^{-1}(\cot(\frac{5\pi}{4}))$

Solution: a) $\cos(\cos^{-1}(\frac{1}{2})) = \frac{1}{2}$

b) $\tan^{-1}(\tan(\frac{\pi}{6})) = \frac{\pi}{6}$

c) Be careful with this problem. If you say $\cot^{-1}(\cot(\frac{5\pi}{4})) = \frac{5\pi}{4}$, it is WRONG. Remember $\frac{5\pi}{4}$ is NOT within the interval $[0, \pi]$.
Here is the right answer.
Since $\cot(\frac{5\pi}{4}) = 1$, $\cot^{-1}(\cot(\frac{5\pi}{4})) = \cot^{-1}(1) = \frac{\pi}{4}$.

Example 6: Evaluate each of the following.

a) $\sin(\cos^{-1}(\frac{\sqrt{3}}{2}))$ b) $\tan^{-1}(\cot(\frac{5\pi}{6}))$ c) $\csc^{-1}(\sec(\pi))$

Solution: a) Let $\theta = \cos^{-1}(\frac{\sqrt{3}}{2})$. This means $\cos(\theta) = \frac{\sqrt{3}}{2}$. So, $\theta = \frac{\pi}{6}$.
$\sin(\cos^{-1}(\frac{\sqrt{3}}{2})) = \sin(\frac{\pi}{6}) = \boxed{\frac{1}{2}}$

b) $\cot(\frac{5\pi}{6}) = \frac{\cos(\frac{5\pi}{6})}{\sin(\frac{5\pi}{6})} = \frac{\frac{-\sqrt{3}}{2}}{\frac{1}{2}} = -\sqrt{3}$.
$\tan^{-1}(\cot(\frac{5\pi}{6})) = \tan^{-1}(-\sqrt{3}) = \boxed{\frac{-\pi}{3}}$

c) $\sec(\pi) = \frac{1}{\cos(\pi)} = \frac{1}{-1} = -1$. So, $\csc^{-1}(\sec(\pi)) = \csc^{-1}(-1) = \boxed{\frac{-\pi}{2}}$

45

Example 7: Find each of the following.

a) $\sec(\sin^{-1}(\frac{-3}{4}))$ b) $\sin(\tan^{-1}(x))$ c) $\cos(\csc^{-1}(x))$

Solution: a) Let $\theta = \sin^{-1}(\frac{-3}{4})$. Then $\sin(\theta) = \frac{-3}{4}$, and θ is in quadrant IV.

Here is the sketch of the angle.

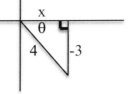

Using the Pythagorean Theorem,
we have $4^2 = (-3)^2 + x^2$
$$\Rightarrow x = \sqrt{16 - 9} = \sqrt{7}$$

So, $\sec(\sin^{-1}(\frac{-3}{4})) = \sec(\theta) = \frac{hypotenuse}{adjacent} = \frac{4}{\sqrt{7}}$

b) Let $\theta = \tan^{-1}(x)$. This means $\tan(\theta) = x = \frac{x}{1}$.

So, if you draw a right triangle with the acute angle θ, then x is the opposite side, and 1 is the adjacent side.

Here it is.

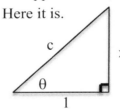

Using the Pythagorean Theorem,
we have $c^2 = 1^2 + x^2$
$$\Rightarrow c = \sqrt{1 + x^2}$$

So, $\sin(\tan^{-1}(x)) = \sin(\theta) = \frac{x}{\sqrt{1 + x^2}}$

c) Let $\theta = \csc^{-1}(x)$. This means $\csc(\theta) = x = \frac{x}{1}$.

Now draw the triangle.

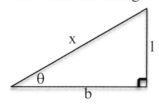

Using the Pythagorean Theorem,
we have $x^2 = 1^2 + b^2$
$$\Rightarrow b^2 = x^2 - 1$$
$$\Rightarrow b = \sqrt{x^2 - 1}$$

So, $\cos(\csc^{-1}(x)) = \cos(\theta) = \frac{\sqrt{x^2 - 1}}{x}$

Practice 6

Calculate each of the following.

1) $\sin^{-1}(0)$

2) $\cos^{-1}(0)$

3) $\sin^{-1}(-1)$

4) $\tan^{-1}(\frac{-1}{\sqrt{3}})$

5) $\cot^{-1}(\frac{-1}{\sqrt{3}})$

6) $\sec^{-1}(-\sqrt{3})$

7) $\tan(\csc^{-1}(-2))$

8) $\sin^{-1}(\sin(\frac{7\pi}{6}))$

9) $\cos^{-1}(\cos(2\pi))$

10) $\cot^{-1}(\cot(480°))$

11) $\tan(\sec^{-1}(x))$

12) $\csc(\cos^{-1}(x))$

Graph each of the following.

13) $y = \sin^{-1}(x)$

14) $y = \cos^{-1}(x)$

15) $y = \cot^{-1}(x)$

16) $y = 2\sin^{-1}(x) + 1$

Use a calculator to find the radian measure of each angle. Round your answers to two decimal places.

17) $\cos^{-1}(0.37)$

18) $\sin^{-1}(-0.62)$

19) $\tan^{-1}(4.0)$

20) $\sec^{-1}(3.2)$

Bonus: Simplify.

$$\sin^{-1}(-1) + 2\cos^{-1}(-1) + 3\tan^{-1}(-1) - 4\cot^{-1}(-1) - 5\sec^{-1}(-1) - 6\csc^{-1}(-1)$$

Did you know?

To multiply a number by **25**, you can just divide that number by 4 and attach two zeros at the end. For example, 84 x 25 = **2100** because 84 ÷ 4 = 21.

Here is another example. 216 x 25 = **5400** because 216 ÷ 4 = 54.

Lesson 7: Trigonometric Identities

Based on the definitions of the six Trigonometric functions from lesson 1, you can easily establish the following two identities:

1) Reciprocal Identities

$$\sin x = \frac{1}{\csc x} \qquad\qquad \csc x = \frac{1}{\sin x}$$

$$\cos x = \frac{1}{\sec x} \qquad\qquad \sec x = \frac{1}{\cos x}$$

$$\tan x = \frac{1}{\cot x} \qquad\qquad \cot x = \frac{1}{\tan x}$$

2) Quotient Identities

$$\tan x = \frac{\sin x}{\cos x} \qquad\qquad \cot x = \frac{\cos x}{\sin x}$$

With a little further observation from the unit circle, you can see the following identities:

3) Even-Odd Identities

$$\sin(-x) = -\sin x \qquad\qquad \csc(-x) = -\csc(x)$$

$$\cos(-x) = \cos x \qquad\qquad \sec(-x) = \sec(x)$$

$$\tan(-x) = -\tan x \qquad\qquad \cot(-x) = -\cot(x)$$

Note that only cosine and secant are even functions; the other four are odd.

In case you forgot, an even function is a function such that $f(-x) = f(x)$. For example, $f(x) = x^2$ is even because $f(4) = (4)^2 = 16$ and $f(-4) = (-4)^2 = 16$. In general, $f(-x) = (-x)^2 = x^2 = f(x)$.

An odd function is a function such that $f(-x) = -f(x)$. For example, $f(x) = x^3$ is odd because $f(2) = (2)^3 = 8$ and $f(-2) = (-2)^3 = -8$. In general, $f(-x) = (-x)^3 = -x^3 = -f(x)$.

Finally, by looking at a right triangle, you can prove the following two identities:

4) Cofunction Identities

$$\sin(\tfrac{\pi}{2} - x) = \cos(x) \qquad\qquad \cos(\tfrac{\pi}{2} - x) = \sin(x)$$

$$\tan(\tfrac{\pi}{2} - x) = \cot(x) \qquad\qquad \cot(\tfrac{\pi}{2} - x) = \tan(x)$$

$$\sec(\tfrac{\pi}{2} - x) = \csc(x) \qquad\qquad \csc(\tfrac{\pi}{2} - x) = \sec(x)$$

5) Pythagorean Identities

$$\sin^2(x) + \cos^2(x) = 1$$

$$\tan^2(x) + 1 = \sec^2(x)$$

$$\cot^2(x) + 1 = \csc^2(x)$$

First, let's talk about the cofunction identities. Why is $\sin(\tfrac{\pi}{2} - x) = \cos(x)$? Well, if you have a right triangle, the two acute angles always add to 90°. For example, if one angle is 40°, then the other angle must be 50°. (See figure)

$$\sin(40°) = \frac{a}{c} \qquad \cos(50°) = \frac{a}{c}$$

$$\sin(90° - 50°) = \cos(50°)$$

Next, let's see why $\sin^2(x) + \cos^2(x) = 1$.

Consider the right triangle shown here.

Since $\qquad \sin(x) = \dfrac{a}{c} \qquad$ and $\qquad \cos(x) = \dfrac{b}{c}$

we have $\qquad \sin^2(x) = \left(\dfrac{a}{c}\right)^2 \qquad$ and $\qquad \cos^2(x) = \left(\dfrac{b}{c}\right)^2$.

$$\sin^2(x) + \cos^2(x) = \frac{a^2}{c^2} + \frac{b^2}{c^2} = \frac{a^2 + b^2}{c^2}$$

but in a right triangle, $a^2 + b^2 = c^2$; so $\dfrac{a^2 + b^2}{c^2} = \dfrac{c^2}{c^2} = 1$.

This proves $\sin^2(x) + \cos^2(x) = 1$. Note that it does not matter what the angle is, sine square of an angle plus cosine square of the **same** angle, equals 1.
For example, $\sin^2(40°) + \cos^2(40°) = 1$, but $\sin^2(40°) + \cos^2(47°) \neq 1$.

By the way, in Trigonometry, $\sin^2(x)$ means $(\sin x)^2$. $\sin^2(x)$ is NOT the same as $\sin(x^2)$ or $\sin x^2$.

You can use $\sin^2(x) + \cos^2(x) = 1$ to derive the other two identities.

Start with $\quad \sin^2(x) + \cos^2(x) = 1$, and divide both sides by $\cos^2(x)$,

you have $\qquad \dfrac{\sin^2(x)}{\cos^2(x)} + \dfrac{\cos^2(x)}{\cos^2(x)} = \dfrac{1}{\cos^2(x)}$

which is $\qquad \tan^2(x) + 1 = \sec^2(x)$ ✓

Now if you start with $\sin^2(x) + \cos^2(x) = 1$, and divide both sides by $\sin^2(x)$,

you have $\qquad \dfrac{\sin^2(x)}{\sin^2(x)} + \dfrac{\cos^2(x)}{\sin^2(x)} = \dfrac{1}{\sin^2(x)}$

which is $\qquad 1 + \cot^2(x) = \csc^2(x)$ ✓

It is very important that you know these identities inside out. In fact, sometimes you need to recognize variations of these identities. For example, knowing $\sin^2(x) + \cos^2(x) = 1$ also means you should know $\sin^2(x) = 1 - \cos^2(x)$, or $\cos^2(x) = 1 - \sin^2(x)$, or even $\cos(x) = \pm\sqrt{1 - \sin^2(x)}$.

Example 1:   Given $\csc(x) = -4$, $\tan(x) > 0$. Find the values of the remaining five Trigonometric functions.

Solution: Using the reciprocal identities, $\sin(x) = \dfrac{1}{\csc x} = \dfrac{-1}{4}$

$$\cos(x) = \pm\sqrt{1 - \sin^2(x)} = \pm\sqrt{1 - \left(\tfrac{-1}{4}\right)^2}$$

$$= \pm\sqrt{1 - \left(\tfrac{1}{16}\right)} = \pm\sqrt{\left(\tfrac{15}{16}\right)} = \pm\dfrac{\sqrt{15}}{4}$$

But since $\sin(x)$ is negative and $\tan(x)$ is positive, x must be in quadrant III. So,

$$\cos(x) = \dfrac{-\sqrt{15}}{4} \quad \Rightarrow \quad \sec x = \dfrac{1}{\cos x} = \dfrac{-4}{\sqrt{15}}$$

$$\tan(x) = \dfrac{\sin x}{\cos x} = \dfrac{\frac{-1}{4}}{\frac{-\sqrt{15}}{4}} = \dfrac{1}{\sqrt{15}} \quad \Rightarrow \quad \cot(x) = \dfrac{1}{\tan x} = \sqrt{15}$$

Example 2:   Simplify $\cot(\tfrac{\pi}{2} - x)\cos x - \sin(-x)$

Solution: Using the cofunction identities, $\cot(\tfrac{\pi}{2} - x) = \tan(x)$. So,

$$\cot(\tfrac{\pi}{2} - x)\cos x - \sin(-x) = \tan x \cos x - \sin(-x)$$

$$= \dfrac{\sin x}{\cos x} \cdot \cos x - (-\sin x)$$

$$= \sin x + \sin x = 2\sin x$$

51

***Example 3*:**   Verify the identity $\dfrac{\tan^2(x) + \sec^2(x) - 1}{2(\sec x + 1)} = \sec x - 1$.

***Solution*:**   Using the Pythagorean Identity, we know $\tan^2(x) + 1 = \sec^2(x)$.

So, $\tan^2(x) = \sec^2(x) - 1$.

Substitute this into the left hand side, we have

$$\frac{\tan^2(x) + \sec^2(x) - 1}{2(\sec x + 1)} = \frac{\sec^2(x) - 1 + \sec^2(x) - 1}{2(\sec x + 1)} = \frac{2\sec^2(x) - 2}{2(\sec x + 1)} =$$

$$\frac{2(\sec^2(x) - 1)}{2(\sec x + 1)} = \frac{(\sec x - 1)(\sec x + 1)}{(\sec x + 1)} = \sec x - 1 \ \checkmark$$

Important: Do not treat an identity to be verified as an equation. The idea is, you don't know for sure $\dfrac{\tan^2(x) + \sec^2(x) - 1}{2(\sec x + 1)} = \sec x - 1$. That is why you need to "verify" it. You need to work with one side and get it equals to the other side.

You cannot move things around as if you are solving equations.

For example, you cannot start with $\dfrac{\tan^2(x) + \sec^2(x) - 1}{2(\sec x + 1)} = \sec x - 1$ and claim that $\tan^2(x) + \sec^2(x) - 1 = 2(\sec x + 1)(\sec x - 1)$ is true.

Multiplying both sides by $2(\sec x + 1)$ means you **assumed** the original problem is true, before you are verifying.

Here is another example. Suppose your friend asked you to verify or check to see if $10 = -9 - 1$ is true. Clearly, we can see that this simple statement is not true.

However, if you start out by assuming it's true and square both sides, you will get $10^2 = (-9 - 1)^2$ or $100 = 100$, which confirms your wrong assumption and lead you to the wrong conclusion.

Example 4:   Verify the identity $\dfrac{1-(\sin x + \cos x)^2}{\sec(-x)\tan(-x)} = 2\cos^3 x$

Solution: Since the left hand side is more complicated, try to simplify it.

$$\frac{1-(\sin x + \cos x)^2}{\sec(-x)\tan(-x)} = 2\cos^3 x$$

$$\frac{1-(\sin^2 x + 2\sin x\cos x + \cos^2 x)}{\sec(-x)\tan(-x)} = 2\cos^3 x$$

$$\frac{1-(1 + 2\sin x\cos x)}{\sec(-x)\tan(-x)} = 2\cos^3 x$$

$$\frac{-2\sin x\cos x}{\sec(-x)\tan(-x)} = 2\cos^3 x$$

$$\frac{-2\sin x\cos x}{-\sec(x)\tan(x)} = 2\cos^3 x$$

$$\frac{2\sin x\cos x}{\frac{1}{\cos x}\cdot\frac{\sin x}{\cos x}} = 2\cos^3 x$$

$$2\sin x\cos x \cdot \frac{\cos^2 x}{\sin x} = 2\cos^3 x$$

$$2\cos^3 x = 2\cos^3 x \quad \checkmark$$

In general, it's better to work on the more complicated side and simplify it.

Converting everything to sine and cosine help as well.

Practice 7

Given the following, find the values of all six Trigonometric functions of x.

1) $\sin(\frac{\pi}{2} - x) = \frac{-1}{2}$, $\csc(x) > 0$

2) $\tan x = \frac{\sqrt{7}}{3}$, $\cos(x) = \frac{3}{4}$

3) $\sec(\frac{\pi}{2} - x) = 3$, $\cos(x) < 0$

4) $\cot(-x) = \frac{1}{4}$, $\sin(-x) > 0$

Simplify.

5) $\dfrac{1 - \sec^2(x)}{\cot(-x)}$

6) $\cot x(\sin x - \tan x) - \cos x - \sin^2 x$

7) $\csc^2 x - (2 - \sec^2 x + \tan^2 x)$

8) $\cos(x - \frac{\pi}{2}) + 2\cos(\frac{\pi}{2} - x)$

Verify each identity.

9) $\dfrac{(\frac{1}{\sin x})^2 - 1}{\tan^2(x)} = \cot^4(x)$

10) $\sec x + \csc x + \tan x + \cot x = \dfrac{1 + \sin x + \cos x}{\sin x \cos x}$

11) $\dfrac{1}{1 - \sin(-x)} - \dfrac{1}{1 + \sin(-x)} = -2\sec x \tan x$

12) $\cos^2(20°) + \cos^2(70°) - \sin^2(50°) = \dfrac{1}{\cot^2(40°) + 1}$

13) $(\tan x + 2)^2 = \dfrac{1 + 4\sin x \cos x + 3\cos^2 x}{1 - \sin^2 x}$

14) $\tan^2\left(\dfrac{\pi}{8}\right) + \cot^2\left(\dfrac{3\pi}{8}\right) = 2\csc^2(\dfrac{3\pi}{8}) - \tan^2\left(\dfrac{3\pi}{10}\right) + \csc^2(\dfrac{\pi}{5}) - 3$

15) $\dfrac{1}{\tan x} - \sec x = \dfrac{-1 + \sin^2 x - \sin(-x)}{\sin(-x)\cos(-x)}$

16) $\tan^{\frac{1}{2}}(x)\cos^4(x) + \tan^{\frac{5}{2}}(x)\cos^4(x) = \tan^{\frac{1}{2}}(x)(1 - \sin^2 x)$

17) $\cot^2 x + 5 = \dfrac{5 - 4\cos^2 x}{1 - \sin^2\left(\frac{\pi}{2} - x\right)}.$

18) $\csc^4(x - \dfrac{\pi}{2}) - \csc^4(x) = \dfrac{1 - \cot^2 x}{\cos^4(x) - \cos^6(x)}$

19) Given $u = 7\sin\theta$, and $0 < \theta < \dfrac{\pi}{2}$, write $\sqrt{49 - u^2}$ in terms of θ.

20) Given $v = 6\cot\theta$, and $0 < \theta < \dfrac{\pi}{2}$, write $\sqrt{4v^2 + 144}$ in terms of θ.

Bonus: Simplify.

$$\sec^2(x) + \csc^2(x) - \sec^2(x)\csc^2(x)$$

Did you know?

You can prove anything in the universe if you violate one of the fundamental laws of math. Here is how you can prove 1 = 2.

Suppose x = 1, then it is true that $x(x - 1) = x^2 - 1$.

$$\Rightarrow \quad x(x - 1) = (x + 1)(x - 1) \qquad \textit{factor } x^2 - 1$$

$$\Rightarrow \quad x = (x + 1) \qquad \textit{cancel } x - 1 \textit{ on both sides}$$

$$\Rightarrow \quad 1 = (1 + 1) \qquad \textit{substitute } x = 1$$

$$\Rightarrow \quad 1 = 2 \qquad \textit{simplify}$$

Can you find the mistake in this proof? Check the answer at the end of the lesson.

Lesson 8: Trigonometric Equations

Unlike verifying identities, when solving equations you are supposed to move things around to isolate the Trigonometric function involved. You then use inverse function to isolate the variable.

Example 1:   Solve $\sin x - \frac{1}{2} = 2 - 4\sin(x)$

Solution: Move sinx to the left and all numbers to the right as follow.

$\sin x - \frac{1}{2} = 2 - 4\sin(x)$	Given
$\sin x - \frac{1}{2} + 4\sin(x) = 2 - 4\sin(x) + 4\sin(x)$	Add 4sinx
$5\sin x - \frac{1}{2} = 2$	Simplify
$5\sin x = 2.5$	Add $\frac{1}{2}$
$\sin x = \frac{1}{2}$	Divide by 5
$x = \frac{\pi}{6}$, or $x = \frac{5\pi}{6}$	From the unit circle

However, sinx repeats every 360° or 2π radian. So, there are infinitely many solutions.

Here is how you describe all possible solutions.

$$x = \frac{\pi}{6} + 2k\pi, \text{ or } x = \frac{5\pi}{6} + 2k\pi, \text{ where k is any integer.}$$

(Remember: k could be negative as well).

Example 2:   Solve $5\cot^2(\frac{\pi}{2} - x) - 1 = 4$

Solution: Isolate cot and convert it into tan.

$5\cot^2(\frac{\pi}{2} - x) - 1 = 4$	Given
$5\cot^2(\frac{\pi}{2} - x) - 1 + 1 = 4 + 1$	Add 1
$5\cot^2(\frac{\pi}{2} - x) = 5$	Simplify
$\cot^2(\frac{\pi}{2} - x) = 1$	Divide by 5
$\cot(\frac{\pi}{2} - x) = \pm 1$	Taking square root
$\tan(x) = \pm 1$	Cofunction identity

$\tan(x) = 1$ means $x = \frac{\pi}{4}, \frac{5\pi}{4}, \frac{9\pi}{4}, \ldots$

$\tan(x) = -1$ means $x = \frac{3\pi}{4}, \frac{7\pi}{4}, \frac{11\pi}{4}, \ldots$

Since the period of tan is π, you can describe all solutions as

followed: $\boxed{x = \frac{\pi}{4} + k\pi, \text{ and } x = \frac{3\pi}{4} + k\pi.}$

Notice you don't have to write $x = \frac{\pi}{4} + 2k\pi$, $x = \frac{3\pi}{4} + 2k\pi$, $x = \frac{5\pi}{4} + 2k\pi$, $x = \frac{7\pi}{4} + 2k\pi$. It is redundant. For example, by writing $x = \frac{\pi}{4} + k\pi$, you already included $\frac{5\pi}{4}$ when $k = 1$.

Example 3:   Solve $2\cos^2x + \sqrt{3}\cos x = 0$, $0 \le x < 2\pi$

Solution: This time the problem only asks for solutions in the interval $[0,2\pi)$. Start by factoring out cosx.

$2\cos^2x + \sqrt{3}\cos x = 0$	Given
$\cos x(2\cos x + \sqrt{3}) = 0$	Factor
$\cos x = 0$ or $(2\cos x + \sqrt{3}) = 0$	Zero product property
$x = \dfrac{\pi}{2}, \dfrac{3\pi}{2}$ or $2\cos x = -\sqrt{3}$	
$\cos x = -\dfrac{\sqrt{3}}{2}$	
$x = \dfrac{5\pi}{6}, \dfrac{7\pi}{6}$	

Example 4:   Solve $\sec^2x + \sec x = 2$

Solution: Subtract 2 on both sides and temporarily let $y = \sec x$. You will then have a quadratic equation in terms of y.

$\sec^2x + \sec x = 2$	Given
$\sec^2x + \sec x - 2 = 0$	Subtract 2
$y^2 + y - 2 = 0$	Let $y = \sec x$
$(y + 2)(y - 1) = 0$	Factor
$y = -2$ or $y = 1$	

$\sec x = -2$ or $\sec x = 1$	Substitute secx back into y
$\cos x = \dfrac{-1}{2}$ or $\cos x = 1$	Reciprocal property
$x = \dfrac{2\pi}{3} + 2k\pi$ or $x = 0 + 2k\pi = 2k\pi$	
$x = \dfrac{4\pi}{3} + 2k\pi$	

Example 5: Solve csc4x = $\sqrt{2}$, $0 \leq x < 2\pi$

Solution: Again we have to list all x in the interval $[0, 2\pi)$. Start by rewriting csc4x in terms of sin4x.

csc4x = $\sqrt{2}$ means sin4x = $\dfrac{1}{\sqrt{2}} = \dfrac{\sqrt{2}}{2}$. So,

$$4x = \frac{\pi}{4} + 2k\pi \quad \text{or} \quad 4x = \frac{3\pi}{4} + 2k\pi$$

$$x = \frac{\frac{\pi}{4} + 2k\pi}{4} \quad \text{or} \quad x = \frac{\frac{3\pi}{4} + 2k\pi}{4}$$

$$x = \frac{\pi}{16} + \frac{k\pi}{2} \quad \text{or} \quad x = \frac{3\pi}{16} + \frac{k\pi}{2}, \; k = 0,1,2...$$

For $k = 0$, $x = \dfrac{\pi}{16}$ or $x = \dfrac{3\pi}{16}$

For $k = 1$, $x = \dfrac{\pi}{16} + \dfrac{\pi}{2} = \dfrac{9\pi}{16}$ or $x = \dfrac{3\pi}{16} + \dfrac{\pi}{2} = \dfrac{11\pi}{16}$

For $k = 2$, $x = \dfrac{\pi}{16} + \dfrac{2\pi}{2} = \dfrac{17\pi}{16}$ or $x = \dfrac{3\pi}{16} + \dfrac{2\pi}{2} = \dfrac{19\pi}{16}$

For $k = 3$, $x = \dfrac{\pi}{16} + \dfrac{3\pi}{2} = \dfrac{25\pi}{16}$ or $x = \dfrac{3\pi}{16} + \dfrac{3\pi}{2} = \dfrac{27\pi}{16}$

As you can see, there are 8 solutions even though we restricted x to the interval $[0,2\pi)$. The most common mistake for this problem is, students stop at $4x = \frac{\pi}{4}$ and $4x = \frac{3\pi}{4}$, thinking if they add another 2π to $\frac{\pi}{4}$ or $\frac{3\pi}{4}$, the answers will be over 2π. However, you are not solving for 4x. You are solving for x. So, it's ok for 4x to go over 2π, as long as when you divide by 4, the answers go back down below 2π.

Example 6: Solve $2\tan^2 x - 4\tan x + 1 = 0$, $0 \le x < 2\pi$. Use a calculator to round to two decimal places.

Solution: Temporarily let $y = \tan x$ and use the quadratic formula.

$2\tan^2 x - 4\tan x + 1 = 0$ Given

$2y^2 - 4y + 1 = 0$ Substitute y

$$y = \frac{-b \pm \sqrt{b^2 - 4ac}}{2a} = \frac{-(-4) \pm \sqrt{(-4)^2 - 4(2)(1)}}{2(2)}$$

$$= \frac{4 \pm \sqrt{16 - 8}}{4} = \frac{4 \pm \sqrt{8}}{4} = \frac{4 \pm 2\sqrt{2}}{4}$$

$$= \frac{2 \pm \sqrt{2}}{2} = 1.71 \text{ or } 0.29$$

So, $\tan x = 1.71$ or $\tan x = 0.29$

$x = \tan^{-1}(1.71)$ or $x = \tan^{-1}(0.29)$

$x = 1.04 + k\pi$ or $x = 0.28 + k\pi$

For $k = 0$, $x = 1.04$ or $x = 0.28$

For $k = 1$, $x = 1.04 + \pi = 4.18$ or $x = 0.28 + \pi = 3.42$

Note: Be sure to change the mode in your calculator from degree to radian, if you have not already done so.

Ok, that's it. It's your turn to practice again. By the way, here is the mistake to the above proof of "$1 = 2$." When you cancel (x-1) on both sides, you are actually dividing both sides by zero because x = 1. Since dividing by zero is undefined, this makes the whole proof invalid.

Practice 8

Solve each equation.

1) $3\tan(x) = \sqrt{3}$

2) $\sqrt{3}\sec(x) + 2 = 0$

3) $\sin^2 x + 2\cos^2 x - 2 = 0$

4) $3\cot^2 x - 1 = 0$

5) $\csc^3 x = \csc x$

6) $3\cot^2 x + \csc^2 x - 5 = 0$

7) $4\sin x + \sin x \cos x = 0$

8) $9\sec^2(\frac{\pi}{2} - x) - 12 = 0$

Find all solutions in the interval $[0, 2\pi)$.

9) $2\sin(-x) + 3\cos(\frac{\pi}{2} - x) = \sin^2 x - 2$

10) $\sin\frac{x}{2} = \frac{1}{2}$

11) $2\cos(3x) + 1 = 0$

12) $\cos\frac{\pi x}{2} = \frac{\sqrt{3}}{2}$

13) $3\cot(2x) = \tan(2x)$

14) $\tan^2 x - \frac{\pi}{2}\tan(-x) = 0$

Use a calculator to find all solutions in the interval $[0, 2\pi)$. Round to two decimal places.

15) $\sin x = -0.7$

16) $\dfrac{1}{1 + \sin x} + \dfrac{1}{1 - \sin x} = 10$

17) $\cot x = \csc^2 x - 7$

18) $2 - 2\sin^2 x = -3\cos x - 1$

19) $3\sin^2 x - 6\sin x + 1 = 0$

20) $\tan^2 x = 4\tan x + 3$

Bonus: Find all solutions in the interval $[0, 2\pi)$.

$$\tan x + \cot x - \sec x \csc x = \sin x + \cos x$$

Did you know?

If you add the first n odd numbers, the sum is n^2. Here are some examples.

$1 + 3 = 4 = 2^2$

$1 + 3 + 5 = 9 = 3^2$

$1 + 3 + 5 + 7 = 16 = 4^2$

$1 + 3 + 5 + 7 + 9 = 25 = 5^2$

You can use this fact to quickly compute the sum of the first 50 odd numbers.

$1 + 3 + 5 + \ldots + 97 + 99 = 50^2 = 2500.$

Lesson 9: Sum and Difference Formulas

In lesson 7, you verified identities to see if they are true. In this lesson, you will learn to use true identities or formulas to solve problems.

There are six important formulas you need to memorize as soon as possible in this lesson. They are called **sum and difference formulas**.

1) $\cos(x + y) = \cos x \cos y - \sin x \sin y$

2) $\cos(x - y) = \cos x \cos y + \sin x \sin y$

3) $\sin(x + y) = \sin x \cos y + \cos x \sin y$

4) $\sin(x - y) = \sin x \cos y - \cos x \sin y$

5) $\tan(x + y) = \dfrac{\tan x + \tan y}{1 - \tan x \tan y}$

6) $\tan(x - y) = \dfrac{\tan x - \tan y}{1 + \tan x \tan y}$

I will show you how to prove these formulas at the end of the lesson. If you are only interested in learning how to use these formulas, and you are not concerned about the proofs because your teacher said you don't need to know them, then you don't have to read the last part. Now, let's see how to use them.

Example 1: Find the exact value of $\sin 75°$.

Solution: Since $75° = 30° + 45°$, $\sin 75° = \sin(30° + 45°)$

$$= \sin 30° \cos 45° + \cos 30° \sin 45°$$

$$= \frac{1}{2} \cdot \frac{\sqrt{2}}{2} + \frac{\sqrt{3}}{2} \cdot \frac{\sqrt{2}}{2}$$

$$= \frac{\sqrt{2}}{4} + \frac{\sqrt{6}}{4} = \frac{\sqrt{2} + \sqrt{6}}{4}$$

Example 2: Find the exact value of $\cos 15°$.

Solution: Since $15° = 45° - 30°$, $\cos 15° = \cos(45° - 30°)$

$$= \cos 45° \cos 30° + \sin 45° \sin 30°$$

$$= \frac{\sqrt{2}}{2} \cdot \frac{\sqrt{3}}{2} + \frac{\sqrt{2}}{2} \cdot \frac{1}{2}$$

$$= \frac{\sqrt{6}}{4} + \frac{\sqrt{2}}{4} = \frac{\sqrt{6} + \sqrt{2}}{4}$$

Notice that the answer is the same as the one in the previous example, because $\sin 75° = \cos 15°$ (cofunction identities, remember?)

Example 3: Find the exact value of $\tan \frac{\pi}{12}$.

Solution: If you can't see how to rewrite $\frac{\pi}{12}$ as a sum or difference, convert it to degree first. However, with practices, eventually you should be able to deal directly with radians.

$$\frac{\pi}{12} = \frac{\pi}{12} \times \frac{180°}{\pi} = 15°.$$

Since $15° = 45° - 30°$, we have $\frac{\pi}{12} = \frac{\pi}{4} - \frac{\pi}{6}$.

So, $\tan\frac{\pi}{12} = \tan(\frac{\pi}{4} - \frac{\pi}{6}) = \dfrac{\tan\frac{\pi}{4} - \tan\frac{\pi}{6}}{1 + \tan\frac{\pi}{4}\tan\frac{\pi}{6}}$

$$= \frac{1 - \frac{1}{\sqrt{3}}}{1 + 1\cdot\frac{1}{\sqrt{3}}} = \frac{\frac{\sqrt{3}}{\sqrt{3}} - \frac{1}{\sqrt{3}}}{\frac{\sqrt{3}}{\sqrt{3}} + \frac{1}{\sqrt{3}}} = \frac{\frac{\sqrt{3} - 1}{\sqrt{3}}}{\frac{\sqrt{3} + 1}{\sqrt{3}}} = \frac{\sqrt{3} - 1}{\sqrt{3} + 1}$$

Example 4: Find the exact value of sin79°cos19° - cos79°sin19°.

Solution: To do this problem, you work backward. Looking at the right hand sides of the six formulas, you can see this fits formula 4. So,

sin79°cos19° - cos79°sin19° = sin(79° - 19°) = sin60° = $\frac{\sqrt{3}}{2}$.

Example 5: Simplify sin(x - π).

Solution: sin(x - π) = sinxcosπ - cosxsinπ

$= $ (sinx)(-1) - (cosx)(0) = -sinx.

Example 6: Solve $\cos(x - \frac{\pi}{3}) - \sin(x - \frac{\pi}{6}) = \frac{-1}{2}$.

Solution: First, $\cos(x - \frac{\pi}{3}) = \cos x\cos\frac{\pi}{3} + \sin x\sin\frac{\pi}{3} = (\cos x)(\frac{1}{2}) + (\sin x)(\frac{\sqrt{3}}{2})$.

Second, $\sin(x - \frac{\pi}{6}) = \sin x\cos\frac{\pi}{6} - \cos x\sin\frac{\pi}{6} = (\sin x)(\frac{\sqrt{3}}{2}) - (\cos x)(\frac{1}{2})$.

So, $\cos(x - \frac{\pi}{3}) - \sin(x - \frac{\pi}{6}) = \frac{-1}{2}$ means

$(\frac{1}{2}\cos x + \frac{\sqrt{3}}{2}\sin x) - (\frac{\sqrt{3}}{2}\sin x - \frac{1}{2}\cos x) = \frac{-1}{2}$.

Simplify gives $\cos x = \frac{-1}{2}$.

Hence, x $= \frac{2\pi}{3} + 2k\pi$ or $\frac{4\pi}{3} + 2k\pi$.

Example 7:   Given $\sin x = \dfrac{-4}{5}$, $\cos y = \dfrac{5}{13}$, x and y are in quadrant IV.

Find $\csc(x + y)$.

Solution: First find $\sin(x + y)$. Then take the reciprocal to get $\csc(x + y)$.

$$\sin(x + y) = \sin x \cos y + \cos x \sin y$$
$$= \dfrac{-4}{5} \cdot \dfrac{5}{13} + \cos x \sin y$$

Since you are given $\sin x = \dfrac{-4}{5}$ and x is in quadrant IV, you can draw the triangle as follow.

By the Pythagorean Theorem, the adjacent side must be 3.

So, $\cos x = \dfrac{3}{5}$.

Similarly, you are given $\cos y = \dfrac{5}{13}$ and y is in quadrant IV, you can draw the triangle as follow.

By the Pythagorean Theorem, the opposite side must be -12.

So, $\sin y = \dfrac{-12}{13}$.

Finally, $\sin(x + y) = \dfrac{-4}{5} \cdot \dfrac{5}{13} + \dfrac{3}{5} \cdot \dfrac{-12}{13} = \dfrac{-20}{65} + \dfrac{-36}{65} = \dfrac{-56}{65}$

$$\Rightarrow \csc(x + y) = \dfrac{-65}{56}$$

That's it. Next I will prove the formulas. (Again, reading this part is optional). I will start with $\cos(x - y)$.

First, draw a unit circle with angles x and y as follows.

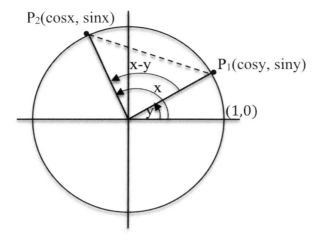

Second, rotate point P_1 and P_2 negative y degree so that P_1 lands on $(1,0)$ and P_2 lands on a new point $P_3(\cos(x - y), \sin(x - y))$.

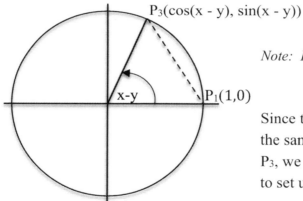

Note: Figure not drawn to scale.

Since the distance from P_1 to P_2 is the same as the distance from P_2 to P_3, we can use the distance formula to set up the following equation.

$$\sqrt{(\cos x - \cos y)^2 + (\sin x - \sin y)^2} = \sqrt{[\cos(x - y) - 1]^2 + [\sin(x - y) - 0]^2}$$

$$\Rightarrow (\cos x - \cos y)^2 + (\sin x - \sin y)^2 = [\cos(x - y) - 1]^2 + [\sin(x - y) - 0]^2$$

$$\Rightarrow \cos^2 x - 2\cos x \cos y + \cos^2 y + \sin^2 x - 2\sin x \sin y + \sin^2 y =$$
$$\cos^2(x - y) - 2\cos(x - y) + 1 + \sin^2(x - y)$$

Since $\sin^2 x + \cos^2 x = 1$, $\sin^2 y + \cos^2 y = 1$, $\sin^2(x - y) + \cos^2(x - y) = 1$,

we have $\quad 1 + 1 - 2\cos x\cos y - 2\sin x\sin y = -2\cos(x - y) + 1 + 1$.

So, $\qquad\qquad -2\cos x\cos y - 2\sin x\sin y = -2\cos(x - y)$.

Hence, $\qquad\qquad \cos(x - y) = \cos x\cos y + \sin x\sin y$ ✓

After proving one formula, the rest follow easily.

To prove $\cos(x + y)$, rewrite it as $\cos(x - (-y))$ and use the above formula, along with the "Even-Odd Identities." Here it is.

$\cos(x + y) = \cos(x - (-y)) = \cos x\cos(-y) + \sin x\sin(-y)$

$$= \cos x\cos y - \sin x\sin y \checkmark$$

To prove $\sin(x + y)$, use the "Cofunction Identities" to rewrite sine in terms of cosine. Here we go.

$\sin(x + y) = \cos[\frac{\pi}{2} - (x + y)] = \cos[(\frac{\pi}{2} - x) - y]$

$$= \cos(\frac{\pi}{2} - x)\cos y + \sin(\frac{\pi}{2} - x)\sin y$$

$$= \sin x\cos y + \cos x\sin y \checkmark$$

$\sin(x - y) = \sin[x + (-y)] = \sin x\cos(-y) + \cos x\sin(-y) = \sin x\cos y - \cos x\sin y$ ✓

$$\tan(x + y) = \frac{\sin(x + y)}{\cos(x + y)} = \frac{\sin x\cos y + \cos x\sin y}{\cos x\cos y - \sin x\sin y}$$

Now divide the numerator and denominator by $\cos x\cos y$. We have,

$$\tan(x + y) = \frac{\dfrac{\sin x\cos y + \cos x\sin y}{\cos x\cos y}}{\dfrac{\cos x\cos y - \sin x\sin y}{\cos x\cos y}} = \frac{\dfrac{\sin x\cos y}{\cos x\cos y} + \dfrac{\cos x\sin y}{\cos x\cos y}}{\dfrac{\cos x\cos y}{\cos x\cos y} - \dfrac{\sin x\sin y}{\cos x\cos y}}$$

$$= \frac{\dfrac{\sin x}{\cos x} + \dfrac{\sin y}{\cos y}}{1 - \dfrac{\sin x\sin y}{\cos x\cos y}} = \frac{\tan x + \tan y}{1 - \tan x\tan y} \checkmark$$

Practice 9

Find the exact value.

1) $\sin 105°$

2) $\cos 255°$

3) $\cos(\frac{35\pi}{18})\cos(\frac{7\pi}{9}) + \sin(\frac{35\pi}{18})\sin(\frac{7\pi}{9})$

4) $\cot(-75°)$

5) $\sin(\frac{22\pi}{21})\cos(\frac{13\pi}{21}) + \cos(\frac{22\pi}{21})\sin(\frac{13\pi}{21})$

6) $\tan\frac{19\pi}{12}$

7) $\sin(\sin^{-1}\frac{2}{3} - \cos^{-1}1)$

8) $\cos(\tan^{-1}\frac{3}{4} - \sin^{-1}\frac{1}{2})$

Use the sum/difference formulas to rewrite each expression.

9) $\sec(x - \frac{3\pi}{2})$

10) $\csc(3\pi - x)$

11) $\tan(\frac{3\pi}{4} - x)$

12) $\cot(\frac{\pi}{3} - x)$

Given $\sin x = \frac{3}{7}$; $\cos y = \frac{-7}{25}$; x and y are in quadrant II. Find each of the following.

13) $\sec(x + y)$

14) $\cot(x - y)$

15) $\tan(y - x)$

16) $\csc(x + y)$

Verify each identity.

17) $\dfrac{\cos(x + y) - \cos(x - y)}{\sin(x + y) + \sin(x - y)} = -\tan y$

18) $\dfrac{\sin(x + y)}{\sin x \sin y} = \cot x + \cot y$

19) Solve the following equation: $\sin(x - \frac{\pi}{4}) - \cos(x - \frac{\pi}{4}) = 1$

20) Use the "Even-Odd Identities" and the formula for $\tan(x + y)$ to prove

$$\tan(x - y) = \frac{\tan x - \tan y}{1 + \tan x \tan y}.$$

Did you know?

Suppose you invested money in the stock market. If you gained 10% during the first year, and you lost 10% during the second year, you would end up with a one-percent loss.

For example, suppose you invested $100. After the first year, you would have $100 plus 10% of 100, which is $110. After the second year, you would have $110 minus 10% of 110, which is $99. So overall you lost $1, which is 1% of your original investment of $100.

Lesson 10: Double-Angle Formulas and Beyond

Before you continue with this lesson, make sure you memorized the sum and difference formulas from the previous lesson. In fact, you can use the sum formulas to derive the **double-angle formulas**.

I. Double-Angle Formulas

1. $\sin(2x) = \sin(x + x) = \sin x \cos x + \cos x \sin x = 2\sin x \cos x$

2. $\cos(2x) = \cos(x + x) = \cos x \cos x - \sin x \sin x = \cos^2 x - \sin^2 x$

3. $\tan(2x) = \tan(x + x) = \dfrac{\tan x + \tan x}{1 - \tan x \tan x} = \dfrac{2\tan x}{1 - \tan^2 x}$

Note: $\sin(2x) \neq 2\sin x$. $\sin(2x) = 2\sin x \cos x$.

Since $\sin^2 x + \cos^2 x = 1$, we can derive two more versions for $\cos(2x)$.

Start with $\quad \cos(2x) = \cos^2 x - \sin^2 x$

$$\cos(2x) = (1 - \sin^2 x) - \sin^2 x = 1 - 2\sin^2 x \ \text{ or}$$

$$\cos(2x) = \cos^2 x - (1 - \cos^2 x) = 2\cos^2 x - 1$$

Using these last two formulas, we can solve for $\sin^2x$ or $\cos^2x$ in terms of $\cos2x$. These are called the **power-reducing formulas**. You will need them in Calculus.

II. Power-Reducing Formulas

1. $\cos(2x) = 1 - 2\sin^2x \Rightarrow 2\sin^2x = 1 - \cos(2x) \Rightarrow \sin^2x = \dfrac{1 - \cos(2x)}{2}$

2. $\cos(2x) = 2\cos^2x - 1 \Rightarrow 2\cos^2x = 1 + \cos(2x) \Rightarrow \cos^2x = \dfrac{1 + \cos(2x)}{2}$

3. $\tan^2x = \dfrac{\sin^2x}{\cos^2x} = \dfrac{1 - \cos(2x)}{2} \div \dfrac{1 + \cos(2x)}{2} = \dfrac{1 - \cos(2x)}{1 + \cos(2x)}$

Finally, if we take the square roots of the power-reducing formulas and let $y = 2x$, so that $x = \dfrac{y}{2}$, we get the **half-angle formulas**.

III. Half-Angle Formulas

1. $\sin^2x = \dfrac{1 - \cos(2x)}{2} \Rightarrow \sin x = \pm\sqrt{\dfrac{1 - \cos(2x)}{2}} \Rightarrow \sin\dfrac{y}{2} = \pm\sqrt{\dfrac{1 - \cos y}{2}}$

2. $\cos^2x = \dfrac{1 + \cos(2x)}{2} \Rightarrow \cos x = \pm\sqrt{\dfrac{1 + \cos(2x)}{2}} \Rightarrow \cos\dfrac{y}{2} = \pm\sqrt{\dfrac{1 + \cos y}{2}}$

3. $\tan^2x = \dfrac{1 - \cos(2x)}{1 + \cos(2x)} \Rightarrow \tan x = \pm\sqrt{\dfrac{1 - \cos(2x)}{1 + \cos(2x)}} \Rightarrow \tan\dfrac{y}{2} = \pm\sqrt{\dfrac{1 - \cos y}{1 + \cos y}}$

We can rationalize the denominator or numerator of $\tan\dfrac{y}{2}$ and get two more versions.

$$\tan\frac{y}{2} = \sqrt{\frac{1-\cos y}{1+\cos y}} = \sqrt{\frac{1-\cos y}{1+\cos y}} \times \frac{\sqrt{1-\cos y}}{\sqrt{1-\cos y}} = \frac{1-\cos y}{\sqrt{1-\cos^2 y}} = \frac{1-\cos y}{\sqrt{\sin^2 y}} = \frac{1-\cos y}{\sin y}$$

or

$$\tan\frac{y}{2} = \sqrt{\frac{1-\cos y}{1+\cos y}} = \sqrt{\frac{1-\cos y}{1+\cos y}} \times \frac{\sqrt{1+\cos y}}{\sqrt{1+\cos y}} = \frac{\sqrt{1-\cos^2 y}}{1+\cos y} = \frac{\sqrt{\sin^2 y}}{1+\cos y} = \frac{\sin y}{1+\cos y}$$

As you can see, just in this lesson alone, there are 13 formulas that you need to memorize. If you remember how one formula can be derive from another, you will have a greater chance of recalling them from memory. It all starts with cos(x - y), then cos(x + y), then sin(x ± y) and tan(x ±y), then the double-angle formulas, then power-reducing formulas, then half-angle formulas.

Example 1: Given $\cos x = \frac{-12}{13}$, $\pi < x < \frac{3\pi}{2}$.

           Calculate. a) sin2x   b) cos2x   c) tan2x

Solution: Since $\cos x = \frac{-12}{13}$ and x is in quadrant III, we can draw the following triangle.

a) $\sin 2x = 2\sin x \cos x$

$$= 2\left(\frac{-5}{13}\right)\left(\frac{-12}{13}\right)$$

$$= \frac{120}{169}$$

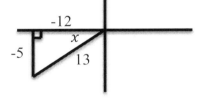

b) $\cos 2x = \cos^2 x - \sin^2 x = \left(\frac{-12}{13}\right)^2 - \left(\frac{-5}{13}\right)^2 = \frac{144-25}{169} = \frac{119}{169}$

c) $\tan 2x = \frac{\sin 2x}{\cos 2x} = \frac{120}{169} \div \frac{119}{169} = \frac{120}{169} \times \frac{169}{119} = \frac{120}{119}$

71

Example 2: Rewrite $\sin^4 x$ in terms of the first power of cosine.

Solution: Use the power-reducing formulas multiple times as follow.

$$\sin^4 x = \sin^2 x \cdot \sin^2 x = \left(\frac{1 - \cos 2x}{2}\right)\left(\frac{1 - \cos 2x}{2}\right)$$

$$= \frac{1 - 2\cos 2x + \cos^2 2x}{4}$$

$$= \frac{1 - 2\cos 2x + \left(\frac{1 + \cos 4x}{2}\right)}{4}$$

$$= \frac{\frac{2}{2} - \frac{4\cos 2x}{2} + \left(\frac{1 + \cos 4x}{2}\right)}{4}$$

$$= \frac{\frac{3 - 4\cos 2x + \cos 4x}{2}}{4}$$

$$= \frac{3 - 4\cos 2x + \cos 4x}{8}$$

$$= \frac{3}{8} - \frac{1}{2}\cos 2x + \frac{1}{8}\cos 4x$$

Example 3: Find the exact values of $\sin 22.5°$, $\cos 22.5°$, $\tan 22.5°$.

Solution: Since $22.5° = \frac{45°}{2}$, use the half-angle formulas.

$$\sin 22.5° = \sin\frac{45°}{2} = +\sqrt{\frac{1 - \cos 45°}{2}} = \sqrt{\frac{1 - \frac{\sqrt{2}}{2}}{2}} = \sqrt{\frac{\frac{2}{2} - \frac{\sqrt{2}}{2}}{2}} =$$

$$\sqrt{\frac{\frac{2 - \sqrt{2}}{2}}{2}} = \sqrt{\frac{2 - \sqrt{2}}{4}} = \frac{\sqrt{2 - \sqrt{2}}}{2}$$

$$\cos 22.5° = \cos\frac{45°}{2} = +\sqrt{\frac{1 + \cos 45°}{2}} = \sqrt{\frac{1 + \frac{\sqrt{2}}{2}}{2}} = \frac{\sqrt{2 + \sqrt{2}}}{2}$$

$$\tan 22.5° = \frac{\sin 22.5°}{\cos 22.5°} = \frac{\sqrt{2 - \sqrt{2}}}{2} \div \frac{\sqrt{2 + \sqrt{2}}}{2} = \frac{\sqrt{2 - \sqrt{2}}}{\sqrt{2 + \sqrt{2}}}$$

Notice that I did not put "±" in front of the radical signs, because 22.5° is in quadrant I, so sin22.5° and cos22.5° should be positive.

Example 4: Find the exact values of $\csc\dfrac{7\pi}{8}$ and $\sec\dfrac{7\pi}{8}$.

Solution: Since $\dfrac{7\pi}{8} = \dfrac{{}^{7\pi}/_{4}}{2}$, use the half-angle formulas for sine and cosine.

$$\sin\frac{7\pi}{8} = \sin\frac{{}^{7\pi}/_{4}}{2} = +\sqrt{\frac{1-\cos\frac{7\pi}{4}}{2}} = \sqrt{\frac{1-\frac{\sqrt{2}}{2}}{2}} = \frac{\sqrt{2-\sqrt{2}}}{2}.$$

$$\csc\frac{7\pi}{8} = \frac{2}{\sqrt{2-\sqrt{2}}}$$

$$\cos\frac{7\pi}{8} = \cos\frac{{}^{7\pi}/_{4}}{2} = -\sqrt{\frac{1+\cos\frac{7\pi}{4}}{2}} = -\sqrt{\frac{1+\frac{\sqrt{2}}{2}}{2}} = -\frac{\sqrt{2+\sqrt{2}}}{2}.$$

$$\sec\frac{7\pi}{8} = -\frac{2}{\sqrt{2+\sqrt{2}}}$$

Notice that $\dfrac{7\pi}{8}$ is in quadrant II, so csc is positive and sec is negative.

Practice 10

Given $\tan x = -\frac{3}{4}, \frac{3\pi}{2} < x < 2\pi$. Find each of the following.

1) $\sin 2x$ 2) $\cos 2x$ 3) $\tan 2x$

4) $\tan 3x$ 5) $\csc 4x$ 6) $\sec(-2x)$

Find the exact value of each of the following.

7) $\sin(112.5°)$ 8) $\tan(-112.5°)$ 9) $\sec(112.5°)$

10) $\cos(\frac{9\pi}{8})$ 11) $\csc(\frac{-9\pi}{8})$ 12) $\cot(\frac{9\pi}{8})$

13) Use the double-angle formulas to simplify $8\sin x \cos x \cos 2x$.

Rewrite each expression in terms of the first power of cosine.

14) $\sin^2 x \cos^2 x$ 15) $\sin^2 x \cos^4 x$

Verify the identity.

16) $\dfrac{2}{1 + \cos 2x} = \sec^2 x$

17) $-\sin x \cos x + 1 - \cos^2 2x + 2.5\sin 2x = \sin 2x(\sin 2x + 2)$

Solve.

18) $\cos 2x = \sin^2 \dfrac{x}{2}$ 19) $3\sin 2x - \sin x = -5\cos(x - \dfrac{\pi}{2})$

20) Use the double-angle formula to rewrite $\csc(2\cos^{-1} x)$.
        (Hint: Let $y = \cos^{-1} x$).

Bonus: Derive a triple angle formula for $\sin 3x$ in terms of powers of $\sin x$.

Did you know?

If you take a multiple-choice test containing 10 questions, and each question has 4 choices, then the probability of you guessing and getting 100% on the test is 1 out of 1,048,576. This is like flipping a fair coin 20 times and seeing 20 heads in a row. It's not going to happen. So study!

Lesson 11: Product-to-Sum and Sum-to-Product

As you can see, in the previous two lessons, there are 19 formulas you need to memorize. There are 8 more formulas in this lesson. However, most likely you don't need to memorize them; your teachers will let you look them up during tests. In case they don't, you should remember how to derive them, again using the sum and difference formulas. We know that

$$\cos(x - y) = \cos x \cos y + \sin x \sin y \qquad \text{and}$$
$$\cos(x + y) = \cos x \cos y - \sin x \sin y$$

Adding the two equations together, we get

$$\cos(x - y) + \cos(x + y) = 2\cos x \cos y. \text{ This means}$$

1) $\qquad \cos x \cos y = \boxed{\dfrac{1}{2}[\cos(x - y) + \cos(x + y)]}$

Now suppose we are subtracting instead of adding the first two equations, we get

$$\cos(x - y) - \cos(x + y) = 2\sin x \sin y. \text{ This means}$$

2) $\qquad \sin x \sin y = \boxed{\dfrac{1}{2}[\cos(x - y) - \cos(x + y)]}$

We also know that

$$\sin(x + y) = \sin x \cos y + \cos x \sin y \qquad \text{and}$$
$$\sin(x - y) = \sin x \cos y - \cos x \sin y$$

Adding the above two equations, we get

$$\sin(x + y) + \sin(x - y) = 2\sin x\cos y. \text{ This means}$$

3) $\qquad \sin x\cos y = \boxed{\dfrac{1}{2}[\sin(x + y) + \sin(x - y)]}$

Now subtract instead of add, we get

$$\sin(x + y) - \sin(x - y) = 2\cos x\sin y. \text{ This means}$$

4) $\qquad \cos x\sin y = \boxed{\dfrac{1}{2}[\sin(x + y) - \sin(x - y)]}$

Example 1: Write each product as a sum or difference.

   a) $\sin 3x \cdot \sin 4x$ b) $\cos(-x) \cdot \sin(-5x)$

Solution: a) Since $\sin x\sin y = \dfrac{1}{2}[\cos(x - y) - \cos(x + y)]$

$$\sin 3x \cdot \sin 4x = \dfrac{1}{2}[\cos(3x - 4x) - \cos(3x + 4x)]$$

$$= \dfrac{1}{2}[\cos(-x) - \cos(7x)]$$

$$= \dfrac{1}{2}[\cos(x) - \cos(7x)].$$

   b) Using the even-odd identities, $\cos(-x) = \cos(x)$ and
    $\sin(-5x) = -\sin(5x)$.

$$\cos(-x) \cdot \sin(-5x) = -\cos(x) \cdot \sin(5x)$$

$$= -\dfrac{1}{2}[\sin(x + 5x) - \sin(x - 5x)]$$

$$= -\dfrac{1}{2}[\sin(6x) - \sin(-4x)]$$

$$= -\dfrac{1}{2}[\sin(6x) + \sin(4x)].$$

Ok let's finish this lesson with the Sum-to-Product Formulas. Here they are.

5) $\qquad \sin x + \sin y = 2\sin(\dfrac{x + y}{2})\cos(\dfrac{x - y}{2})$

6) $\sin x - \sin y = 2\sin(\frac{x-y}{2})\cos(\frac{x+y}{2})$

7) $\cos x + \cos y = 2\cos(\frac{x+y}{2})\cos(\frac{x-y}{2})$

8) $\cos x - \cos y = -2\sin(\frac{x+y}{2})\sin(\frac{x-y}{2})$

You can use the Product-to-Sum formulas to derive the Sum-to-Product formulas. For example, to prove formula (8), we can use formula (2) as follows.

$\sin(a)\sin(b) = \frac{1}{2}[\cos(a - b) - \cos(a + b)]$. This means

$2\sin(a)\sin(b) = \cos(a - b) - \cos(a + b)$.

Now use this substitution trick. Let $x = a - b$ and $y = a + b$. Then

$$x + y = a - b + a + b = 2a, \text{ or } a = \frac{x+y}{2}$$

$$x - y = a - b - (a + b) = -2b \text{ or } b = -\frac{(x-y)}{2}.$$

So, $\cos(a - b) - \cos(a + b) = 2\sin(a)\sin(b)$ means

$$\cos x - \cos y = 2\sin(\frac{x+y}{2}) \sin(\frac{-(x-y)}{2}) = -2\sin(\frac{x+y}{2}) \sin(\frac{(x-y)}{2}) \checkmark$$

You can prove the other 3 formulas in a similar manner.

Example 2:   Write each sum or difference as a product.
       a) $\sin(5x) - \sin(3x)$ b) $\cos(7x) + \cos(4x)$

Solution: a) Using formula (6), we have
       $$\sin(5x) - \sin(3x) = 2\sin(\frac{5x-3x}{2})\cos(\frac{5x+3x}{2}) = 2\sin x\cos 4x.$$

       b) Using formula (7), we have
       $$\cos(7x) + \cos(4x) = 2 \cos\left(\frac{7x+4x}{2}\right) \cos\left(\frac{7x-4x}{2}\right)$$

       $$= 2\cos(\frac{11x}{2})\cos(\frac{3x}{2}).$$

Practice 11

Write each product as a sum or difference.

1) $\sin 8x \sin 3x$

2) $\cos x \sin 5x$

3) $\sin\frac{x}{3}\cos\frac{x}{4}$

4) $\cos(-2x)\cos(\frac{x}{6})$

5) $4\cos\frac{5\pi}{3}\sin\frac{3\pi}{4}$

6) $10\sin(\frac{x}{2} + \frac{y}{2})\sin(\frac{x}{2} - \frac{y}{2})$

Write each sum or difference as a product.

7) $\sin x + \sin 5x$

8) $4\cos 3x + 4\cos 4x$

9) $\dfrac{\sin 6x - \sin 3x}{2}$

10) $\cos\frac{x}{5} - \cos\frac{x}{7}$

11) $\cos(x - \frac{3\pi}{2}) + \cos(\frac{\pi}{2} - x)$

12) $\sin(x - \frac{\pi}{6}) + \sin(x + \frac{\pi}{6})$

Verify each identity.

13) $\dfrac{\sin 10x - \sin 2x}{4\sin 2x \cos 2x} = \cos^2(3x) - \sin^2(3x)$

14) $\left(\dfrac{\cos x + \cos 2x}{\sin x + \sin 2x}\right)^2 = -1 + \csc^2(1.5x)$

15) $\dfrac{\sin 11x - \sin 3x}{\cos 11x + \cos 3x} = \dfrac{2\tan 2x}{1 - \tan^2(2x)}$

16) $\cos 8x - \cos 4x = -8\sin x \cos x \sin 3x \cos 3x$

Solve each equation.

17) $\cos x + \cos 3x = 0$

18) $\sin 5x - \sin x = 0$

19) $2\sin(\frac{3x}{4})\cos(-\frac{x}{4}) = \sin\frac{x}{2} + \frac{7\pi}{6}$

20) $\cos(\frac{\pi x}{3}) - \cos(\frac{\pi x}{4}) = 0$

Bonus: Compute with using a calculator.

$$\sqrt{\cos(100\pi) + \sin(100.5\pi) + \tan(101\pi) + \cot(101.5\pi) + \sec(102\pi) + \csc(102.5\pi)}$$

Lesson 12: Law of Sines

Up to this point, we only use Trigonometry to solve right triangles. What about triangles with no right angle? Well, if you are given two angles and any side (i.e., AAS or ASA) or you are given two sides and an angle opposite one of them (i.e., SSA), you can use the law of sines to solve such triangles.

The Law of Sines states that, given ΔABC as shown below,

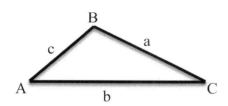

$$\frac{a}{\sin A} = \frac{b}{\sin B} = \frac{c}{\sin C}$$ or

$$\frac{\sin A}{a} = \frac{\sin B}{b} = \frac{\sin C}{c}$$

To prove a part of the law, you can drop an altitude from vertex B to its opposite side as follows.

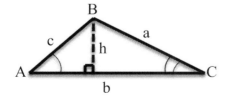

From the figure, we know

$\sin A = \dfrac{h}{c}$ and $\sin C = \dfrac{h}{a}$. This means

$h = c\sin A$ and $h = a\sin C$. So,

$c\sin A = a\sin C$. Therefore, $\dfrac{\sin A}{a} = \dfrac{\sin C}{c}$ ✓

You can finish proving the other parts by drawing the altitude from a different vertex and proceed the same way.

Now let's see how to apply the law when you know two angles and a side (i.e., AAS or ASA).

Example 1: Solve the given triangle.

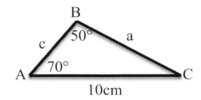

Solution: In any triangle, the sum of the three angles must be 180°.
So, m∠C = 180° - (50 + 70) = $\boxed{60°}$

Also, $\dfrac{a}{sinA} = \dfrac{b}{sinB}$. This means $\dfrac{a}{sin70°} = \dfrac{10}{sin50°}$.

$$So,\ \ a = \frac{10sin70°}{sin50°} = \boxed{12.3\text{cm}}$$

Finally, $\dfrac{c}{sinC} = \dfrac{b}{sinB}$. This means $\dfrac{c}{sin60°} = \dfrac{10}{sin50°}$.

$$So,\ \ c = \frac{10sin60°}{sin50°} = \boxed{11.3\text{cm}}$$

Example 2: Solve the given triangle.

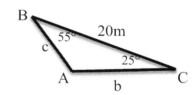

Solution: m∠A = 180° - (25 + 55) = $\boxed{100°}$

Since $\dfrac{b}{sin55°} = \dfrac{20}{sin100°}$,

this means b $= \dfrac{20sin55°}{sin100°} = \boxed{16.6\text{m}}$ and

$$\frac{c}{sin25°} = \frac{20}{sin100°},$$

so c $= \dfrac{20sin25°}{sin100°} = \boxed{8.6\text{m}}$

Next, let's study the SSA case. It turns out that when you are given two sides and an angle, sometimes no triangle can be formed from the given information. Other times, one triangle can be formed; still other times, two triangles are possible. That is why the SSA case is called the **ambiguous case**. Below is an example in which a triangle cannot be created from the given measures.

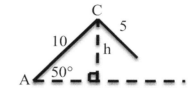

Given m∠A = 50°, a = 5, b = 10.

"a" is not long enough to form a triangle no matter how you adjust angle C.

How do we know? Well, the shortest distance from vertex C to the opposite side is h, and $sin50° = \dfrac{h}{10}$.

$$So, h = 10sin50° = 7.7 > 5.$$

As you can see, the height can tell you how many triangles can be formed. Here are the rules.

1)   if a < h, no triangle

2)   if a = h, one (right) triangle

3)   if a ≥ b, one triangle (as shown below)

4)   if h < a < b, two triangles (as shown below)

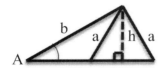

5)   If angle A is obtuse and a ≤ b, then no triangle can be formed. This is because from Geometry, we know that the longest side should be opposite the biggest angle. Since angle A is obtuse, side "a" must be the longest side; it cannot be less than or equal side "b."

6)   If angle A is obtuse and a > b, then exactly one triangle can be formed.

Now, you could try to memorize the six ambiguous cases. However, if you use the Law of Sines correctly, it will tell you how many triangles are possible.

Example 3: Solve the given triangle.

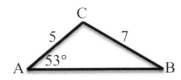

Solution: First, find angle B because b = 5 is known.

$$\frac{\sin 53°}{7} = \frac{\sin B}{5} \Rightarrow \sin B = \frac{5\sin 53°}{7} \Rightarrow \sin B = 0.57.$$

Between 0° and 180°, there are two possible angle B such that $\sin B = 0.57$. One angle is in quadrant I, and the other is in quadrant II.
$B_1 = \sin^{-1}(0.57) = 34.8°$ and
$B_2 = 180 - 34.8 = 145.2°.$

Since A = 53°, B **cannot** equal 145.2°, because A + B is more than 180°. (Remember, in any triangle, the sum of the three angles must equal 180°).

So, B = $\boxed{34.8°}$ and C = 180 - (53° + 34.8°) = $\boxed{92.2°}$

Finally, $\dfrac{c}{\sin 92.2°} = \dfrac{7}{\sin 53°} \Rightarrow c = \dfrac{7\sin 92.2°}{\sin 53°} = \boxed{8.8}$

Example 4: Solve the triangle: a = 9, b = 12, A = 70°.

Solution: $\dfrac{\sin 70°}{9} = \dfrac{\sin B}{12} \Rightarrow \sin B = \dfrac{12\sin 70°}{9} \Rightarrow \sin B = 1.3$

$\Rightarrow$ No solution.

Remember, sine of any angle must always be less than or equals 1.

Example 5: $a = 11, b = 15, A = 40°$.

Solution: $\dfrac{\sin 40°}{11} = \dfrac{\sin B}{15} \Rightarrow \sin B = \dfrac{15 \sin 40°}{11} \Rightarrow \sin B = 0.88$

$\Rightarrow B = 61.2°$ or $B = 180° - 61.2° = 118.8°$.

Since $A = 40°$,

if $B = 61.2°$, then $C = 180 - (40° + 61.2°) = 78.8°$ and

if $B = 118.8°$, then $C = 180 - (40° + 118.8°) = 21.2°$

If $C = 78.8°$, then $\dfrac{c}{\sin 78.8°} = \dfrac{11}{\sin 40°} \Rightarrow c = \dfrac{11 \sin 78.8°}{\sin 40°} = 16.8$

If $C = 21.2°$, then $\dfrac{c}{\sin 21.2°} = \dfrac{11}{\sin 40°} \Rightarrow c = \dfrac{11 \sin 21.2°}{\sin 40°} = 6.2$

Here is the summary of the two possible solutions:

$a = 11, b = 15, A = 40°, B = 61.2°, C = 78.8°, c = 16.8$

or $a = 11, b = 15, A = 40°, B = 118.8°, C = 21.2°, c = 6.2$

Notice that the given information stay the same in both triangles.

Ok, that's it. Here is how you can solve the above average-speed problem. Suppose the distance from your house to work is 120 miles (you can pick any number. I choose 120 because it is a multiple of 40 and 60). In the morning, it takes you 3 hours to go to work. In the evening, it takes you 2 hours to go home. Since you takes 5 hours to drive a total distance of 240 miles, your average rate is 240 miles divided by 5 hours = 48 mile per hour.

Practice 12

Solve each triangle. Round to the nearest tenth.

1)

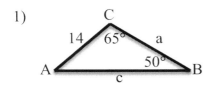

2)

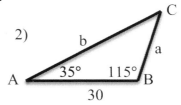

3) B = 40°, b = 9, C = 70°

4) a = 20, b = 25, A = 48°

5) b = 17, c = 22, B = 95°

6) a = 18, b = 13, A = 33°

7) a = 11, c = 16, C = 65°

8) b = 60, c = 72, C = 110°

9) A = 52°, C = 74°, b = 80

10) A = 102°, a = 30, c = 19

Without solving, determine how many triangle can be formed from the given information.

11) a = 5, b = 15, A = 20°

12) b = 18, c = 22, B = 47°

13) a = 51, c = 58, A = 64°

14) b = 9, c = 16, C = 36°

15) a = 12, b = 10, B = 109°

16) a = 29 , c = 32 , A = 58°

17) b = 5, c = 8, B = 30°

18) c = 34, b = 40, C = 93°

19) Suppose an airplane flies 200 miles East before it makes a 50° angle turn and flies for another 30 minutes. If the airplane is 250 miles from the initial position, how far did it fly during the 30-minute? (See figure).

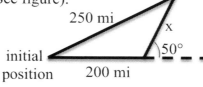

20) Suppose you stand on top of a building and look down. You see a person on the ground that is 25 feet from you and 10 feet from his dog. (See figure). If the angle of depression from you to him is 70°, and the angle of depression from you to the dog is 50°, how far are you from the dog?

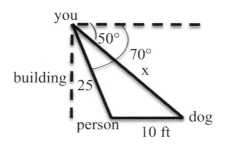

Bonus: Use the law of sines and the fact in a triangle m∠C = 180 – (A + B) to show that c = a(cosB) + b(cosA).

Did you know?

You can predict a person's phone number by using a math trick. Here is how.

Ask that person to take out a calculator and do the following (without showing you the answers):

1) Enter the first three digits of the phone number.

2) Multiply by 13, plus 9, multiply by 7, multiply by 11, minus 911, plus 218,

3) Minus the first three digits of the phone number, multiply by 20,

4) Plus the last four digits of the phone number, plus the last four digits again, then divide by 2.

5) Now ask the person to show you the calculator. The phone number is on the display.

Lesson 13: Law of Cosines

In the previous lesson, you learned to use the Law of Sines to solve AAS, ASA and SSA problems. To solve triangles with two sides and the included angle given (SAS) or three sides given (SSS), you can use the Law of Cosines.

Given triangle ABC as shown below, the law of cosines state that

$a^2 = b^2 + c^2 - 2bc\cos A$, or

$b^2 = a^2 + c^2 - 2ac\cos B$, or

$c^2 = a^2 + b^2 - 2ab\cos C$

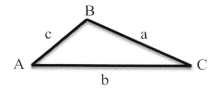

Note that you only need to memorize 1 formula, not 3. In words, the law of cosines say that if you squared any one side of a triangle, it is equals the sum of the squares of the other 2 sides minus two times their product multiply by the cosine of their included angles.

Also notice how similar these formulas are compared to the Pythagorean Theorem. In fact, the last formula is Pythagorean Theorem with $C = 90°$, because $\cos 90° = 0$.

To prove one of the formulas, place the above triangle in a coordinate plane as shown.

Since $\sin A = \dfrac{y}{c}$ and $\cos A = \dfrac{x}{c}$,

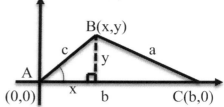

we have $y = c\sin A$ and $x = c\cos A$.

Since "a" equals the distance from vertex B to vertex C,

$$a = \sqrt{(x - b)^2 + (y - 0)^2} \Rightarrow a^2 = (x - b)^2 + (y)^2$$

$$\Rightarrow a^2 = (c\cos A - b)^2 + (c\sin A)^2$$

$$\Rightarrow a^2 = c^2\cos^2 A - 2bc\cos A + b^2 + c^2\sin^2 A$$

$$\Rightarrow a^2 = b^2 + c^2\sin^2 A + c^2\cos^2 A - 2bc\cos A$$

$$\Rightarrow a^2 = b^2 + c^2(\sin^2 A + \cos^2 A) - 2bc\cos A$$

$$\Rightarrow a^2 = b^2 + c^2 - 2bc\cos A \ \checkmark \qquad (\text{because } \sin^2 A + \cos^2 A = 1)$$

Of course you can prove the other 2 formulas the same way.

Example 1:   Solve the given triangle.

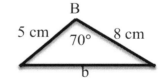

Solution: First, "b" is missing, so put "b" on the left.

$$b^2 = a^2 + c^2 - 2ac\cos B \Rightarrow b^2 = 8^2 + 5^2 - 2(8)(5)\cos 70°$$
$$\Rightarrow b^2 = 64 + 25 - 80\cos 70°$$
$$\Rightarrow b^2 = 89 - 27.4 = 61.6$$
$$\Rightarrow \boxed{b = 7.8 \text{ cm}}$$

To find angle A, you can use the law of sines or the law of cosines. However, if you use the law of sines, sometimes you need to check the ambiguous cases. So, I am going to use the law of cosines.

$$a^2 = b^2 + c^2 - 2bc\cos A \implies \cos A = \frac{a^2 - b^2 - c^2}{-2bc}$$

$$\implies \cos A = \frac{64 - 61.6 - 25}{-2(7.8)(5)}$$

$$\implies \cos A = \frac{-22.6}{-78} = 0.2897$$

$$\implies A = \cos^{-1}(0.2897) = \boxed{73.2°}$$

$$C = 180 - (70° + 73.2°) = \boxed{36.8°}$$

Example 2: Solve the given triangle: a = 10 ft, b = 14 ft, c = 21 ft.

Solution: Since 3 sides are given, you can solve for any one of the angles first.

$$a^2 = b^2 + c^2 - 2bc\cos A \implies \cos A = \frac{a^2 - b^2 - c^2}{-2bc}$$

$$\implies \cos A = \frac{10^2 - 14^2 - 21^2}{-2(14)(21)}$$

$$\implies \cos A = \frac{100 - 196 - 441}{-2(14)(21)}$$

$$\implies \cos A = \frac{-537}{-588} = 0.9133$$

$$\implies A = \cos^{-1}(0.9133) = \boxed{24.0°}$$

$$\text{Similarly, } \cos B = \frac{b^2 - a^2 - c^2}{-2ac} \implies \cos B = \frac{14^2 - 10^2 - 21^2}{-2(10)(21)}$$

$$\implies \cos B = \frac{196 - 100 - 441}{-2(10)(21)}$$

$$\implies \cos B = \frac{-345}{-420} = 0.8214$$

$$\implies B = \cos^{-1}(0.8214) = \boxed{34.8°}$$

$$C = 180 - (24.0° + 34.8°) = \boxed{121.2°}$$

I will now switch gear and talk a little bit about the term **bearing**.
In Trigonometry, we often work with angles that the terminal side makes with the x-axis, except when it involves "bearing" problems. "Bearing" deals with the acute angle a ray makes with the y-axis or the North-South line.

Here are some examples.

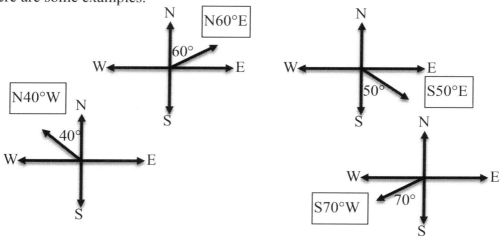

So, to describe the bearing, mention North or South first, follow by the acute angle measure, then West or East.

Example 3: An airplane flies 100 miles West from point A to point B, before it changes direction and flies another 120 miles toward point C. (See figure). If the airplane is 180 miles from the point of departure, what is the bearing of the airplane from point B to C?

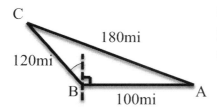

Solution: Use the law of cosines to find angle B.

$$b^2 = a^2 + c^2 - 2ac\cos B \Rightarrow \cos B = \frac{b^2 - a^2 - c^2}{-2ac}$$

$$\Rightarrow \cos B = \frac{180^2 - 120^2 - 100^2}{-2(120)(100)} = \frac{1}{3}$$

$$\Rightarrow \quad B = 109.47°$$

Since the question asks for bearing, you have to subtract 90° from angle B (see picture if you are not sure why).

109.47° - 90° = 19.47°. So the bearing is N19.47°W

Practice 13

Solve each triangle. Round to the nearest tenth.

1)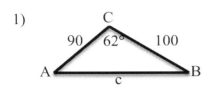

2)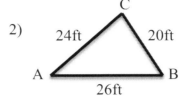

3) a = 4, b = 7, c = 10

4) A = 65°, b = 6, c = 9

5) B = 107°, a = 15, c = 20

6) C = 40°, a = 5, b = 8

7) a = 12, b = 3, c = 14

8) a = 11, b = 11, c = 7

9) A = 35°30'18", b = 16, c = 22

10) C = 90°12'9", a = 1, b = 2

Find the measure of the angle x.

11)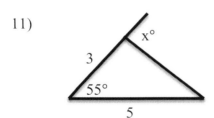

12)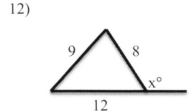

13)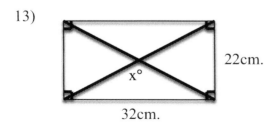

14)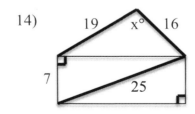

15) Find the value of y.

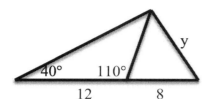

16)

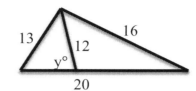

17) Two ships leave a port at 5 P.M.. One travels at a bearing of N62°E at 25 miles per hour; the other travels at 30 miles per hour with a bearing of S55°E. How far apart are the ships at 7 P.M.? (Round to the nearest tenth).

18) An ant on a wall runs 2 feet North from point X to point Y; it changes direction and runs 3 feet toward point Z. (See figure). If the ant ends up 4 feet away from its starting point, what is the bearing from point Y to point Z?

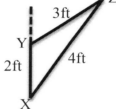

19) From City A, City B is 200 miles away and the bearing is N40°E. From City A, City C is 220 miles away and the bearing is N70°E. What is the distance between City B and City C?

20) The distance between 3 schools in a district is shown. Find the bearing of the second school from the third school.

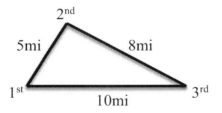

Did you know?

Here is how you can predict a person's birthday. Ask that person to take out a calculator and do the following (without showing you the answers):

1) Enter the birth month.
2) Plus 1, multiply by 25, minus 2, multiply by 4, plus 3,
3) Plus the birth date, multiply by 50, minus 5, multiply by 2,
4) Plus the two-digit birth year, minus 9490.
5) Now ask the person to show you the calculator. The birthday is on the display.

Lesson 14: Areas of Triangles

In Geometry you learned that the area of a triangle is $A = \frac{1}{2} \times$ base $\times$ height.

But, what if you don't know the height, and are given SAS instead? Well, it turns out you can easily use Trigonometry to still find the area.

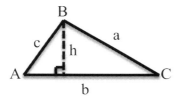

From the figure, we have

$$\sin A = \frac{h}{c} \Rightarrow h = c \cdot \sin A$$

$$\text{Area} = \frac{1}{2}bh = \boxed{\frac{1}{2}bc\sin A}$$

Similarly, if you are given sides a, b and angle C, then Area $= \boxed{\frac{1}{2}ab\sin C}$

or,   if you are given sides a, c and angle B, then Area $= \boxed{\frac{1}{2}ac\sin B}$

Example 1: Find the area of the triangle.

Solution: Since you are given 2 sides and the included angle,

$$\text{Area} = \tfrac{1}{2} \times 5 \times 9 \times sin80° = \boxed{22.2}$$

Now, what if you know 3 sides (SSS)? Can you find the area of the triangle? Well, you can use the law of cosines to find an angle, and then apply the above area formulas. Alternatively, you can use **Heron's Formula**, which states that the area of a triangle with sides a, b, c is

$$\boxed{\text{Area} = \sqrt{s(s - a)(s - b)(s - c)}, \text{ where } s = \frac{a + b + c}{2}}$$

You can use the law of cosines to prove Heron's Formula. However, your teacher will not require you to prove it. So, we will not prove it here. Using the formula is also quite easy. Here is an example.

Example 2:   Find the area of a triangle given a = 10, b = 13, c = 19.

Solution: First calculate the value of s.

$$s = \frac{a + b + c}{2} = \frac{10 + 13 + 19}{2} = \frac{42}{2} = 21$$

$$\text{Area} = \sqrt{21(21 - 10)(21 - 13)(21 - 19)} = \sqrt{21 \cdot 11 \cdot 8 \cdot 2}$$

$$= \sqrt{3696} = 60.8$$

Example 3:   Find the area of the shaded region.

Solution: In lesson 2, you learned that the area of a sector in radian is $\frac{\theta}{2}r^2$.

So, first convert 70° to radian.

$$70° \times \frac{\pi}{180°} = \frac{7\pi}{18}$$

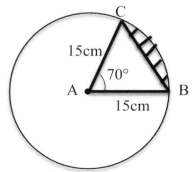

Area of a sector is

$$\frac{1}{2} \times \frac{7\pi}{18} \times 15^2 = \frac{1575\pi}{36} \approx 137.38$$

Area of ΔABC is $\frac{1}{2}(15)(15)\sin 70° \approx 105.72$

Area of shaded region = 137.38 - 105.72 = $\boxed{31.66 \text{ cm}^2}$

Example 4: Given area of ΔABC equals 153.2 m^2, find the area of ΔBCD as shown.

Solution: Let x be each of the four congruent sides.

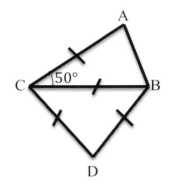

Since the area of ΔABC = 153.2, we have $153.2 = \frac{1}{2} \cdot x \cdot x \cdot \sin 50°$

$\Rightarrow \quad 306.4 = x^2 \cdot \sin 50°$

$\Rightarrow \quad \frac{306.4}{\sin 50°} = x^2$

$\Rightarrow \quad x = 20$

Since ΔBCD is equilateral, all three sides equal 20 m. Using Herons' formula, the area of ΔBCD is

$$\sqrt{30(30 - 20)(30 - 20)(30 - 20)},$$

because $s = \frac{a + b + c}{2} = \frac{20 + 20 + 20}{2} = \frac{60}{2} = 30.$

So, the area $= \sqrt{30(10)(10)(10)} = \sqrt{30000} = \boxed{173.2 \ m^2}$

Practice 14

Find the area of each triangle. Round to the nearest tenth.

1) a = 40, b = 45, C = 55°

2) a = 12, c = 6, B = 32°

3) b = 7, c = 11, A = 63°

4) a = 45, b = 45, C = 120°

5) a = 4, b = 8, c = 11

6) a = 15, b = 13, c = 15

7) A = 50°, B = 75°, c = 10

8) A = 45°, C = 55°, a = 40

9) a = $\sqrt{3}$, b = $\sqrt{12}$, C = 60°

10) A = 30°, C = 98°, c = $\sqrt{10}$

11) Find the radius of the circle given the area of ΔABC equals 100 ft^2.

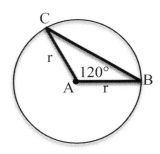

12) Find the area of the rhombus shown.

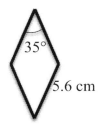

13) Find the area of the parallelogram shown.

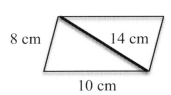

14) Find the value of x, given the area A = 50.

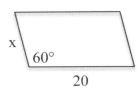

15) Given a triangle with area A = $\sqrt{20}$, perimeter P = 10, one side a = 3, find the two missing sides. (Hint: Use Heron's formula).

16) Show that the area of the triangle shown below is A = 10asinx.

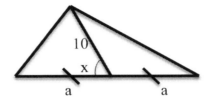

17) Use the answer from problem 16) to find the area of the triangle when a = 6 and x = 40°.

18) Use the answer from problem 16) to find "a," given A = 100 and x = 70°.

19) Use the answer from problem 16) to find x, given A = 40 and a = 5.

20) Find the area of ΔABC.

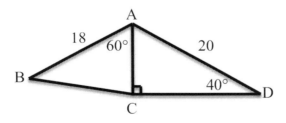

Lesson 15: Polar Coordinates

Up to now in math, you use the rectangular coordinate system to locate a point. For instance, to describe the location of point P as shown here, you tell people to move 4 units to the right and 3 units up from the origin. However, this is not the only way to do thing.

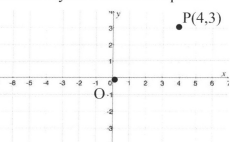

To plot point P, you could also tell people the length of segment OP, along with the angle that OP makes with the x-axis. So instead of using (x, y), you use **(r, θ)** to describe a point in **polar coordinates** as shown below.

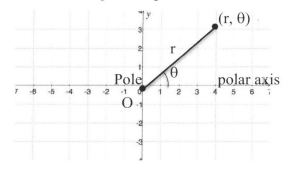

"r" is the radius.

The x-axis is the "polar axis."

The origin is the "Pole."

Example 1: Plot the following points in polar coordinates.

     a) $(3, \frac{\pi}{4})$ b) $(2, \frac{5\pi}{6})$ c) $(3, \frac{-7\pi}{4})$

Solution: a) Since $\theta = \frac{\pi}{4}$, first draw a ray that makes a 45-degree angle with the polar axis. Then, go out 3 units as shown.

b) Since $\theta = \frac{5\pi}{6}$, first draw a ray that makes a 150-degree angle with the polar axis. Then, go out 2 units as shown.

c) Since $\theta = \frac{-7\pi}{4}$, rotate clockwise 315 degree. Then, draw a ray and go out 3 units as shown.

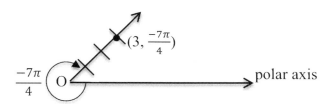

If you look carefully, you will notice that the points in examples 1a and 1c describe the exact same location. That is the biggest difference between rectangular versus polar coordinates. In polar coordinates, there are infinitely many ways to describe a point. Not only $(3, \frac{\pi}{4}) = (3, \frac{-7\pi}{4})$, $(3, \frac{\pi}{4}) = (3, \frac{9\pi}{4}) = (3, \frac{17\pi}{4})$ . . . In fact, every time you add or subtract 2π, you will land on the same point. So, **$(r, \theta) = (r, \theta \pm 2\pi)$**.

Also, you can use negative values for r. If r < 0, you can plot the point by going to angle θ and come out positive r units, then jump or reflect across. Here is an example of how to plot $(-2, \frac{\pi}{3})$.

1) Go to $\frac{\pi}{3}$

2) Draw $(2, \frac{\pi}{3})$

3) Reflect across to get $(-2, \frac{\pi}{3})$.

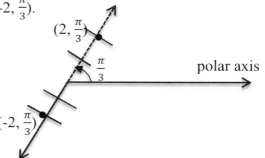

Notice that $(-2, \frac{\pi}{3})$ can also be described as $(2, \frac{4\pi}{3})$, because if you add 180 degree or π to $(2, \frac{\pi}{3})$, you will land on the same spot as $(-2, \frac{\pi}{3})$. In general,

(-r, θ) = (r, θ ± π) = (r, θ ± 3π) = (r, θ ± 5π) = (r, θ ± nπ), where n is odd.

Example 2: Find three points (r, θ) that are equivalent to $(5, \frac{-\pi}{6})$, where $-2\pi < θ < 2\pi$.

Solution: You can add/subtract 2π and keep the "5," or you can add/subtract π and change the "5" to negative 5.

$$(5, \frac{-\pi}{6}) = (5, \frac{-\pi}{6} + 2\pi) = \boxed{(5, \frac{11\pi}{6})}$$

$$(5, \frac{-\pi}{6}) = (-5, \frac{-\pi}{6} + \pi) = \boxed{(-5, \frac{5\pi}{6})}$$

$$(5, \frac{-\pi}{6}) = (-5, \frac{-\pi}{6} - \pi) = \boxed{(-5, \frac{-7\pi}{6})}$$

Next, how do we convert back and forth between the two systems? As it turns out, this is quite simple.

Take a look at the figure below.

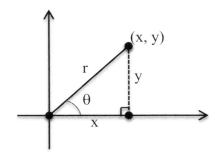

You can see that

$$\cos\theta = \frac{x}{r} \text{ and } \sin\theta = \frac{y}{r}.$$

So, $\boxed{x = r\cos\theta \text{ and } y = r\sin\theta}$

Also, $\boxed{x^2 + y^2 = r^2 \text{ and } \tan\theta = \frac{y}{x}}$

You can use the first two formulas to go from polar to rectangular coordinates, and the last two to go back.

Example 3:   Convert to rectangular coordinates.

        a) $(4, \frac{3\pi}{4})$ b) $(-2, \frac{\pi}{2})$

Solution: a) $x = r\cos\theta \implies x = 4\cos\frac{3\pi}{4} = 4 \cdot \left(-\frac{\sqrt{2}}{2}\right) = -2\sqrt{2}$

                 $y = r\sin\theta \implies y = 4\sin\frac{3\pi}{4} = 4 \cdot \left(\frac{\sqrt{2}}{2}\right) = 2\sqrt{2}$

                 So, the point is (x, y) = $(-2\sqrt{2}, 2\sqrt{2})$.

               b) $x = r\cos\theta \implies$ x = $-2\cos\frac{\pi}{2} = -2 \cdot (0) = 0$

                 $y = r\sin\theta \implies$ y = $-2\sin\frac{\pi}{2} = -2 \cdot (1) = -2$

                 So, the point is (x, y) = (0, -2).

Example 4:   Convert to polar coordinates.

        a) (4, 3)   b) (-5, -5)

Solution: a) $x^2 + y^2 = r^2 \implies 4^2 + 3^2 = r^2 \implies 25 = r^2 \implies r = 5$

               $\tan\theta = \frac{y}{x} \implies \tan\theta = \frac{3}{4} \implies \theta = \tan^{-1}\left(\frac{3}{4}\right)$

Using a calculator, $\theta \approx 37°$ or $\theta = 37 \times \dfrac{\pi}{180} = \dfrac{37\pi}{180}$

So, the point is $(r, \theta) = (5, \dfrac{37\pi}{180})$.

b) $x^2 + y^2 = r^2 \Rightarrow (-5)^2 + (-5)^2 = r^2$

$$\Rightarrow 50 = r^2 \Rightarrow r = 5\sqrt{2}$$

$$tan\theta = \dfrac{y}{x} \Rightarrow tan\theta = \dfrac{-5}{-5} = 1 \Rightarrow \theta = tan^{-1}(1) = \dfrac{5\pi}{4}.$$

So, the point is $(r, \theta) = (5\sqrt{2}, \dfrac{5\pi}{4})$.

Example 5:   Convert each polar equation to rectangular.

   a) $r = 3$ b) $r = 2cos\theta$ c) $\theta = \dfrac{\pi}{6}$

Solution: a) Since $x^2 + y^2 = r^2$, we have $\boxed{x^2 + y^2 = 9}$

   b) Since $r = 2cos\theta,$ multiply both sides by r gives

   $r^2 = 2rcos\theta.$ Substitute $x^2 + y^2$ into r^2 and $x = rcos\theta,$

   we have $\boxed{x^2 + y^2 = 2x}$

   c) Since $tan\theta = \dfrac{y}{x},$ $tan\dfrac{\pi}{6} = \dfrac{y}{x}$

   $$\Rightarrow \dfrac{1}{\sqrt{3}} = \dfrac{y}{x} \Rightarrow \boxed{y = \dfrac{\sqrt{3}}{3}x}$$

Example 6:   Convert $y = x^2 + 2x$ to polar equation.

Solution: Substitute $y = rsin\theta$ and $x = rcos\theta$ into the equation, we get

   $$rsin\theta = (rcos\theta)^2 + 2(rcos\theta)$$

$\Rightarrow$ $r\sin\theta = r^2\cos^2\theta + 2r\cos\theta$

$\Rightarrow$ $\sin\theta = r\cos^2\theta + 2\cos\theta$

$\Rightarrow$ $\sin\theta - 2\cos\theta = r\cos^2\theta$

$\Rightarrow$ $\dfrac{\sin\theta - 2\cos\theta}{\cos^2\theta} = r$

$\Rightarrow$ $r = \dfrac{\sin\theta}{\cos^2\theta} - \dfrac{2\cos\theta}{\cos^2\theta} = \dfrac{\sin\theta}{\cos\theta} \cdot \dfrac{1}{\cos\theta} - \dfrac{2}{\cos\theta}$

$\Rightarrow$ $r = \tan\theta \cdot \sec\theta - 2\sec\theta.$

Practice 15

Plot each point in polar coordinates.

1) $(1, \pi)$ 2) $(2, \frac{3\pi}{2})$ 3) $(5, \frac{-3\pi}{4})$

4) $(-3, \frac{-5\pi}{3})$ 5) $(-4, 3.25\text{rad})$ 6) $(0, \frac{\pi}{6})$

Find three equivalent points (r, θ) for each given, where $-2\pi < \theta < 2\pi$.

7) $(1, \frac{\pi}{3})$ 8) $(2, \frac{-\pi}{4})$

9) $(-3, \frac{5\pi}{6})$ 10) $(1.5, 2\text{rad})$

Convert to rectangular coordinates.

11) $(5, \frac{5\pi}{4})$ 12) $(10, \frac{11\pi}{6})$

13) $(-1, \frac{\pi}{3})$ 14) $(2, 3\text{rad})$

Convert to polar coordinates.

15) $(1, 2)$ 16) $(6, -8)$

17 $(-5, 12)$ 18) $(-\sqrt{3}, -\sqrt{2})$

19) Convert $r = 6\sin\theta + 8\cos\theta$ to rectangular equation.

20) Convert $y = x^2$ to polar equation.

Bonus: Convert $x^2 - xy = y^3 - y^2$ to polar equation.

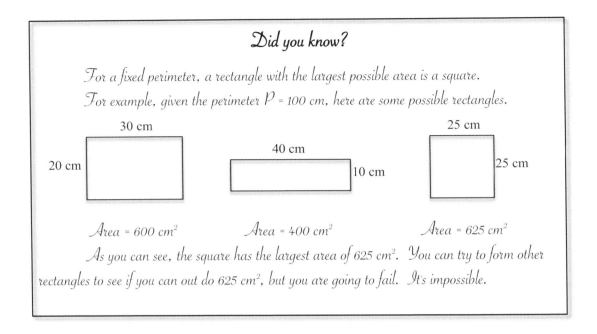

Lesson 16: Graphing Polar Equations

To graph rectangular equations, you use graph paper or rectangular grids. To graph polar equations, you will use circular grids.

Example 1:   Graph $r = 2\sin\theta$.

Solution: When you are new to any graphs, you should first try to plot points.

Here is the table of values for θ, where $0 \le \theta \le \pi$.

θ	0	$\frac{\pi}{6}$	$\frac{\pi}{4}$	$\frac{\pi}{3}$	$\frac{\pi}{2}$	$\frac{2\pi}{3}$	$\frac{3\pi}{4}$	$\frac{5\pi}{6}$	π
r	0	1	$\sqrt{2}$	$\sqrt{3}$	2	$\sqrt{3}$	$\sqrt{2}$	1	0

Below is the graph.

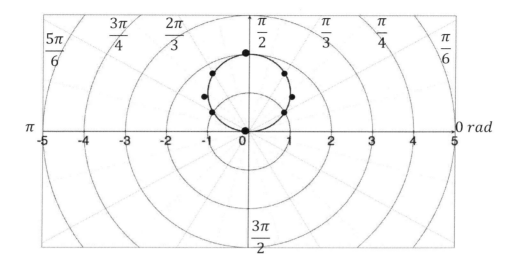

As you can see, the graph is a circle of radius 1. The center is at the point (x, y) = (0, 1). Also, by picking θ from 0 to π, we can trace out the whole graph. In fact, if you did not know that and you go pass π, let's say you picked $\frac{7\pi}{6}$, you will get $r = 2\sin\frac{7\pi}{6} = 2(-\frac{1}{2}) = -1$. But $(-1, \frac{7\pi}{6})$ is the same point as $(1, \frac{\pi}{6})$. So, choosing points beyond π does not add anything to the graph. Lastly, notice that the graph is symmetric with respect to the y-axis. If we knew this, we could have stopped picking points after $\frac{\pi}{2}$ and reflect the graph over the y-axis.

Example 2: Graph $r = 4\cos\theta$.

***S**olution*:   Again let's pick θ values between 0 and π, and see what will happen this time.

Here is the table of values for θ, where $0 \le \theta \le \pi$.

θ	0	$\frac{\pi}{6}$	$\frac{\pi}{4}$	$\frac{\pi}{3}$	$\frac{\pi}{2}$	$\frac{2\pi}{3}$	$\frac{3\pi}{4}$	$\frac{5\pi}{6}$	π
r	4	$2\sqrt{3}$	$2\sqrt{2}$	2	0	-2	$-2\sqrt{2}$	$-2\sqrt{3}$	-4

Notice that after $\frac{\pi}{2}$, r is negative. So, those points will be reflected from the second quadrant to the fourth quadrant. Thus, if we plot the points in the above table, we will get a circle with radius 2 as shown. Also, the graph is symmetric with respect to the x-axis or the polar axis.

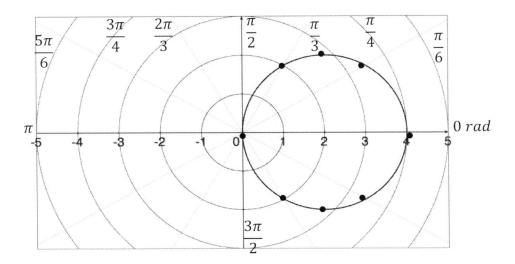

Next time you can test for symmetry before graphing by using the following rules:

1) If you replace θ by π - θ, and you get an equivalent equation, then the graph is symmetric with respect to the y-axis.

2) If you replace θ by -θ, and you get an equivalent equation, then the graph is symmetric with respect to the x-axis or the polar axis.

3) If you replace r by -r, and you get an equivalent equation, then the graph is symmetric with respect to the origin or the pole.

Example 3:   Graph r = 2 + 2sinθ.

Solution: This time let's check for symmetry before graphing.

1) replace θ by π - θ: r = 2 + 2sin(π - θ)
                              r = 2 + 2(sinπcosθ - cosπsinθ)
                              r = 2 + 2(0cosθ - -1sinθ)
                              r = 2 + 2sinθ   ✓   passed

2) replace θ by -θ: r = 2 + 2sin(-θ)
             r = 2 - 2sin(θ)   failed

3) replace r by -r: -r = 2 + 2sin(θ)
             r = -2 - 2sin(θ)   failed

So, the graph is symmetric with respect to the y-axis. This means we should choose values of θ that fall in quadrant I and IV, and then reflect the graph over quadrant II and III.

Here is the table of values.

θ	$\frac{-\pi}{2}$	$\frac{-\pi}{3}$	$\frac{-\pi}{4}$	$\frac{-\pi}{6}$	0	$\frac{\pi}{6}$	$\frac{\pi}{4}$	$\frac{\pi}{3}$	$\frac{\pi}{2}$
r	0	$2-\sqrt{3}$	$2-\sqrt{2}$	1	2	3	$2+\sqrt{2}$	$2+\sqrt{3}$	4

Plotting these points give the right half of the graph below.

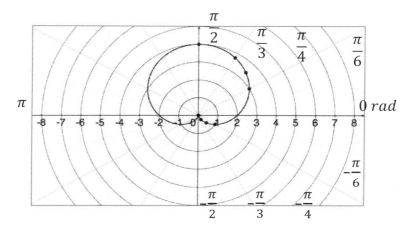

This heart-shape graph is called the "cardioid."
In general, if you have r = a ± asinθ or r = a ± acosθ, then it is a "cardioid."

Example 4:   Graph r = 3 - 2cosθ.

Solution: First check for symmetry.

1) replace θ by π - θ: r = 3 - 2cos(π - θ)
                 r = 3 - 2(cosπcosθ + sinπsinθ)
                 r = 3 - 2(-1cosθ + 0sinθ)
                 r = 3 + 2cosθ   failed

107

2) replace θ by $-\theta$: $r = 3 - 2\cos(-\theta)$

$r = 3 - 2\cos\theta$ ✓ passed

3) replace r by -r: $-r = 3 - 2\cos(\theta)$

$r = -3 + 2\cos(\theta)$ failed

The graph is symmetric with respect to the polar axis. However, because the other two tests failed, the graph may or may not be symmetric with respect to the y-axis and the origin. Note that a "failed" test does not mean the graph is for sure not symmetric, whereas a "passed" result does mean the graph is symmetric.

Since you now get a hang of graphing polar equations, I will pick less points. Here is the table of values.

θ	0	$\dfrac{\pi}{3}$	$\dfrac{\pi}{2}$	$\dfrac{2\pi}{3}$	π
r	1	2	3	4	5

Plotting these points give the top half of the graph below. This graph is called a **limacon**.

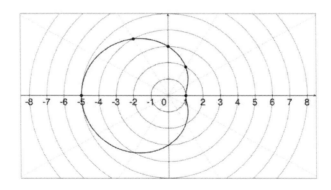

Example 5: Graph $r = 1 - 2\sin\theta$.

Solution: Using our experience with graphing polar equations up to this point, we can see that this graph is symmetric with respect to the y-axis (because it involves sine instead of cosine).

Here is the table of values.

θ	$\dfrac{-\pi}{2}$	$\dfrac{-\pi}{3}$	$\dfrac{-\pi}{6}$	0	$\dfrac{\pi}{6}$	$\dfrac{\pi}{3}$	$\dfrac{\pi}{2}$
r	3	$1+\sqrt{3}$	2	1	0	$1-\sqrt{3}$	-1

Plotting these points give the right half as well as the left inner loop of the graph below. We refer to this graph as a **limacon with inner loop**.

In general, if you have $r = a \pm b\sin\theta$ or $r = a \pm b\cos\theta$, then it is a limacon. Furthermore, if $a < b$, then it has an inner loop.

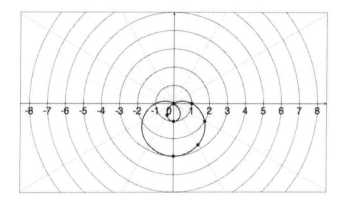

Example 6:   Graph $r = 3\cos2\theta$.

Solution: Since this problem involves cosine, the graph is symmetric with respect to the polar axis.

Also, if you replace θ with $\pi - \theta$, you get
$$r = 3\cos2(\pi - \theta)$$
$$= 3\cos(2\pi - 2\theta)$$
$$= 3\cos(-2\theta)$$
$$= 3\cos(2\theta)$$

So, the graph is symmetric with respect to the y-axis as well. Since the graph is symmetric with respect to both the polar and y-axis, it must be symmetric with respect to the pole or the origin.

Here is the table of values.

θ	0	$\dfrac{\pi}{6}$	$\dfrac{\pi}{4}$	$\dfrac{\pi}{3}$	$\dfrac{\pi}{2}$
r	3	1.5	0	-1.5	-3

109

Plotting these points and reflect them over the polar axis, the y-axis, and the pole, give the **rose curve** below.

In general, if you have $r = a\cos(n\theta)$ or $r = a\sin(n\theta)$, then it is a rose curve. If n is odd, you have n petals. If n is even, you have 2n petals. In this example, n = 2 is even, so you have 4 petals.

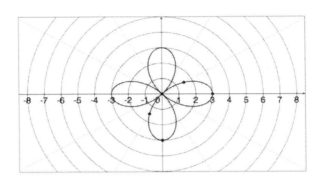

Example 7:   Graph $r^2 = 4\cos2\theta$.

<u>Solution</u>:   Replacing r by -r gives $(-r)^2 = 4\cos2\theta \Rightarrow r^2 = 4\cos2\theta$.
So, this graph is symmetric with respect to the pole or origin. Also, as in example 6, you can easily show this graph is symmetric with respect to the polar axis and the y-axis as well. Finally, since you have r^2 on the left, $4\cos2\theta$ must be positive. This means we should choose θ such that $-\frac{\pi}{4} \le \theta \le \frac{\pi}{4}$, because this is equivalent to $-\frac{\pi}{2} \le 2\theta \le \frac{\pi}{2}$ and cosine is positive in the interval $[-\frac{\pi}{2}, \frac{\pi}{2}]$.
Let's say if you pick $\theta = \frac{\pi}{3}$, you would get
$r^2 = 4\cos(2 \cdot \frac{\pi}{3}) = 4 \cdot \left(\frac{-1}{2}\right) = -2$, which is not possible.

Here is the table of values.

θ	$-\dfrac{\pi}{4}$	$-\dfrac{\pi}{6}$	0	$\dfrac{\pi}{6}$	$\dfrac{\pi}{4}$
r	0	$\pm\sqrt{2}$	±2	$\pm\sqrt{2}$	0

Plotting these points give a **lemniscate**. In general, if you have $r^2 = a^2\cos2\theta$ or $r^2 = a^2\sin2\theta$, then it is a lemniscate. The graph looks like the infinity sign.

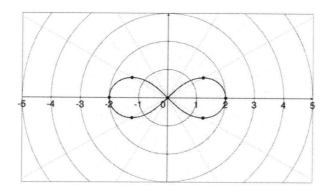

The last thing I want to mention here is that sometimes you can easily graph a polar equation by converting it into rectangular. For example, to graph $\theta = \frac{\pi}{3}$, you use $tan\theta = \frac{y}{x}$. This means $tan\frac{\pi}{3} = \frac{y}{x}$. So, $\sqrt{3} = \frac{y}{x}$ or $y = \sqrt{3}x$. The graph is a line (see below). Similarly, to graph $r = 2$, you use $r^2 = x^2 + y^2$. This means $4 = x^2 + y^2$. The graph is a circle of radius 2.

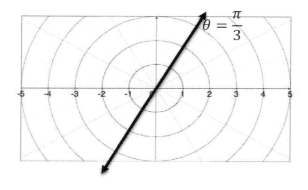

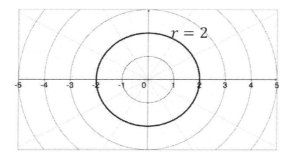

Practice 16

Without graphing, determine whether the following represent a circle, cardioid, limacon, limacon with inner loop, rose curve, lemniscate or straight line.

1) $r = 3 + 5\cos\theta$

2) $r = 2\cos\theta$

3) $\theta = \dfrac{3\pi}{4}$

4) $r = 4 - 4\sin\theta$

5) $r = 7 - 6\sin\theta$

6) $r = 4\cos5\theta$

7) $r^2 = 16\sin2\theta$

8) $r = \dfrac{1}{10\sin\theta - 13\cos\theta}$

Graph each of the following.

9) $r = -3\sin\theta$

10) $r = -\cos\theta$

11) $r = 3 + 3\cos\theta$

12) $r = 1 - 2\cos\theta$

13) $r = 2\cos3\theta$

14) $r = 2\sin2\theta$

15) $r^2 = 9\sin2\theta$

16) $r = 4 + 3\sin\theta$

Convert each polar equation to rectangular and graph.

17) $r = 4$

18) $\theta = \dfrac{\pi}{8}$

19) $r = \dfrac{10}{5\cos\theta + 2\sin\theta}$

20) $r = \dfrac{4}{1 - \sin\theta}$

Bonus: Convert $r = \pm\sqrt{\cot\theta\csc\theta}$ to rectangular equation and graph.

Lesson 17: Complex Number in Polar Form

Recall from Algebra II that a complex number is a number that can be written as a + bi, where "a" is the real part and "bi" is the imaginary part. Just as you can plot a real number on a number line, you can plot a complex number on a complex plane. The x-axis is the real axis, and the y-axis is the imaginary axis.

To plot 2 + 3i, go right 2 units and up 3 units. Similarly, -3 + i means left 3 and up 1; -2i means down 2; 5 means right 5

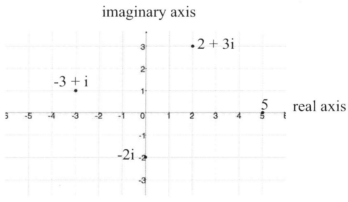

As with real number, absolute value means "distance from zero."

For example, $|-5| = 5$,

and $\qquad |2 + 3i| = \sqrt{(2 - 0)^2 + (3 - 0)^2} = \sqrt{2^2 + 3^2} = \sqrt{13}.$

In general, $\quad |a + bi| = \sqrt{a^2 + b^2}.$

As it turns out, writing complex numbers in polar form allow you to compute powers and roots a lot faster. For example, if you have to compute $(1 - \sqrt{3}i)^{10}$ in Algebra II, that would be painful. However, if you convert $1 - \sqrt{3}i$ to polar form first, the problem becomes simple.

Given a complex number $z = a + bi$, "a" is the "x" component and "b" is the "y" component in the complex plane. So, $a = x = rcos\theta$; $b = y = rsin\theta$.
This means $z = a + bi = rcos\theta + (rsin\theta)i$

$$\Rightarrow \quad z = r(cos\,\theta + isin\,\theta).$$ This is the polar form.

Example 1: Plot the point $z = \sqrt{3} + i$ and convert it to polar form.

Solution: Go right $\sqrt{3}$ unit and up 1 unit as shown below.

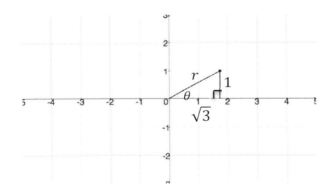

$$r^2 = (\sqrt{3})^2 + 1^2 = 4$$
$$\Rightarrow r = 2.$$
Also, $tan\theta = \dfrac{1}{\sqrt{3}}$
$$\Rightarrow \theta = \dfrac{\pi}{6}.$$
So, $z = r(cos\theta + isin\theta)$

$$= \boxed{2(cos\dfrac{\pi}{6} + isin\dfrac{\pi}{6})}$$

Now let's say you have two complex numbers in polar form.

$z_1 = r_1(cos\theta_1 + isin\theta_1)$ and $z_2 = r_2(cos\theta_2 + isin\theta_2)$.

What is $z_1 \times z_2$?

$z_1 \times z_2 = r_1(cos\theta_1 + isin\theta_1) \times r_2(cos\theta_2 + isin\theta_2)$

$= r_1r_2(cos\theta_1cos\theta_2 + icos\theta_1sin\theta_2 + isin\theta_1cos\theta_2 + i^2sin\theta_1sin\theta_2)$

$= r_1r_2[(cos\theta_1cos\theta_2 - sin\theta_1sin\theta_2) + i(sin\theta_1cos\theta_2 + cos\theta_1sin\theta_2)]$

$= r_1r_2[cos(\theta_1 + \theta_2) + isin(\theta_1 + \theta_2)]$

So, to multiply two complex numbers in polar form, all you have to do is multiply their radii and add their angles.

What about $z_1 \div z_2$?

$$\frac{z_1}{z_2} = \frac{r_1(cos\theta_1 + isin\theta_1)}{r_2(cos\theta_2 + isin\theta_2)} = \frac{r_1(cos\theta_1 + isin\theta_1)}{r_2(cos\theta_2 + isin\theta_2)} \cdot \frac{(cos\theta_2 - isin\theta_2)}{(cos\theta_2 - isin\theta_2)} =$$

$$\frac{r_1(cos\theta_1 cos\theta_2 - icos\theta_1 sin\theta_2 + isin\theta_1 cos\theta_2 - i^2 sin\theta_1 sin\theta_2)}{r_2(cos^2\theta_2 - i^2 sin^2\theta_2)} =$$

$$\frac{r_1[(cos\theta_1 cos\theta_2 + sin\theta_1 sin\theta_2) + i(sin\theta_1 cos\theta_2 - cos\theta_1 sin\theta_2)]}{r_2(cos^2\theta_2 + sin^2\theta_2)} =$$

$$\frac{r_1}{r_2}[cos(\theta_1 - \theta_2) + isin(\theta_1 - \theta_2)].$$

So, to divide two complex numbers in polar form, all you have to do is divide their radii and subtract their angles.

Example 2:   Given $z_1 = 10(cos\frac{3\pi}{4} + isin\frac{3\pi}{4})$, $z_2 = 5(cos\frac{7\pi}{6} + isin\frac{7\pi}{6})$.

Find   a) $z_1 \times z_2$ b) $z_1 \div z_2$

Solution: a) $z_1 \times z_2 = 10 \times 5[cos(\frac{3\pi}{4} + \frac{7\pi}{6}) + isin(\frac{3\pi}{4} + \frac{7\pi}{6})]$

$$= 50(cos\frac{23\pi}{12} + isin\frac{23\pi}{12}).$$

b) $\frac{z_1}{z_2} = \frac{10}{5}[cos(\frac{3\pi}{4} - \frac{7\pi}{6}) + isin(\frac{3\pi}{4} - \frac{7\pi}{6})]$

$$= 2[cos(-\frac{5\pi}{12}) + isin(-\frac{5\pi}{12})]$$

$$= 2[cos(\frac{5\pi}{12}) - isin(\frac{5\pi}{12})].$$

To raise a complex number to an exponent, we can use the multiplication rule repeatedly.

Given $z = r(cos\theta + isin\theta)$

We have $z^2 = r(cos\theta + isin\theta) \times r(cos\theta + isin\theta) = r^2(cos2\theta + isin2\theta)$

$z^3 = r^2(cos2\theta + isin2\theta) \times r(cos\theta + isin\theta) = r^3(cos3\theta + isin3\theta)$

$z^4 = r^3(cos3\theta + isin3\theta) \times r(cos\theta + isin\theta) = r^4(cos4\theta + isin4\theta)$

. . . and so on

In general, $\boldsymbol{z^n = r^n(cosn\theta + isinn\theta)}$. This is called the DeMoivre's Theorem.

Example 3:   Simplify $(1 - \sqrt{3}i)^{10}$.

Solution: First convert $(1 - \sqrt{3}i)$ to polar form.

$r^2 = a^2 + b^2 = 1^2 + (-\sqrt{3})^2 = 1 + 3 = 4 \qquad \Rightarrow \quad r = 2.$

Also, $tan\theta = -\dfrac{\sqrt{3}}{1} \Rightarrow \theta = -\dfrac{\pi}{3}.$

So, $(1 - \sqrt{3}i) = 2(cos\dfrac{-\pi}{3} + isin\dfrac{-\pi}{3}).$

This means $(1 - \sqrt{3}i)^{10} = [2(cos\dfrac{-\pi}{3} + isin\dfrac{-\pi}{3})]^{10}$

$= 2^{10}(cos\dfrac{-10\pi}{3} + isin\dfrac{-10\pi}{3})$

$= 1024(cos\dfrac{10\pi}{3} - isin\dfrac{10\pi}{3})$

$= 1024(-\dfrac{1}{2} - \dfrac{-\sqrt{3}}{2}i)$

$= 1024(-\dfrac{1}{2} + \dfrac{\sqrt{3}}{2}i)$

$= \boxed{-512 + 512\sqrt{3}i}$

Finally, to take the nth root of a complex number $z = r(cos\theta + isin\theta)$, use the following formula.

$$z_k = \sqrt[n]{r}\left[cos\left(\frac{\theta}{n} + \frac{2k\pi}{n}\right) + isin\left(\frac{\theta}{n} + \frac{2k\pi}{n}\right)\right], \quad k = 0, 1, 2, \ldots, n - 1.$$

Example 4:   Find the cube roots of $2 + 2i$.

Solution: First convert $2 + 2i$ to polar form.

$$r^2 = a^2 + b^2 = 2^2 + 2^2 = 4 + 4 = 8 \implies r = 2\sqrt{2}.$$

Also, $tan\theta = \frac{2}{2} \implies \theta = \frac{\pi}{4} = 45°$.

So, $2 + 2i = 2\sqrt{2}(cos45° + isin45°)$.

Since n = 3, k = 0, 1, and 2.

For k = 0, $z_0 = \sqrt[3]{2\sqrt{2}}\left[cos\left(\frac{45°}{3} + \frac{360°}{3} \cdot \mathbf{0}\right) + isin\left(\frac{45°}{3} + \frac{360°}{3} \cdot \mathbf{0}\right)\right]$

$$= \sqrt[3]{2\sqrt{2}}\,(cos15° + isin15°) = \boxed{1.366 + 0.366i}$$

For k = 1, $z_1 = \sqrt[3]{2\sqrt{2}}\left[cos\left(\frac{45°}{3} + \frac{360°}{3} \cdot \mathbf{1}\right) + isin\left(\frac{45°}{3} + \frac{360°}{3} \cdot \mathbf{1}\right)\right]$

$$= \sqrt[3]{2\sqrt{2}}\,(cos135° + isin135°) = \boxed{-1 + i}$$

For k = 2, $z_2 = \sqrt[3]{2\sqrt{2}}\left[cos\left(\frac{45°}{3} + \frac{360°}{3} \cdot \mathbf{2}\right) + isin\left(\frac{45°}{3} + \frac{360°}{3} \cdot \mathbf{2}\right)\right]$

$$= \sqrt[3]{2\sqrt{2}}\,(cos255° + isin255°) = \boxed{-0.366 - 1.366i}$$

Example 5:   Solve the equation $x^5 = 1$.

Solution: According to the Fundamental Theorem of Algebra, a 5th degree equation should have 5 solutions. To solve this problem, take the fifth root on both sides. We have $x = \sqrt[5]{1}$. To get all 5 solutions, first convert $1 = 1 + 0i$ to polar form.

117

Clearly, $r = 1$, and $tan\theta = \frac{0}{1}$. So, $\theta = 0$ rad $= 0°$.

This means $1 + 0i = 1(\cos 0° + i\sin 0°)$

Since n = 5, k = 0, 1, 2, 3, and 4.

For k = 0, $z_0 = 1[\cos(0° + \frac{360°}{5} \cdot \mathbf{0}) + i\sin(0° + \frac{360°}{5} \cdot \mathbf{0})]$

$$= 1(\cos 0° + i\sin 0°) = 1 + 0i = \boxed{1}$$

For k = 1, $z_1 = 1[\cos(0° + \frac{360°}{5} \cdot \mathbf{1}) + i\sin(0° + \frac{360°}{5} \cdot \mathbf{1})]$

$$= 1(\cos 72° + i\sin 72°) = \boxed{0.309 + 0.951i}$$

For k = 2, $z_2 = 1[\cos(0° + \frac{360°}{5} \cdot \mathbf{2}) + i\sin(0° + \frac{360°}{5} \cdot \mathbf{2})]$

$$= 1(\cos 144° + i\sin 144°) = \boxed{-0.809 + 0.588i}$$

For k = 3, $z_3 = 1[\cos(0° + \frac{360°}{5} \cdot \mathbf{3}) + i\sin(0° + \frac{360°}{5} \cdot \mathbf{3})]$

$$= 1(\cos 216° + i\sin 216°) = \boxed{-0.809 - 0.588i}$$

For k = 4, $z_4 = 1[\cos(0° + \frac{360°}{5} \cdot \mathbf{4}) + i\sin(0° + \frac{360°}{5} \cdot \mathbf{4})]$

$$= 1(\cos 288° + i\sin 288°) = \boxed{0.309 - 0.951i}$$

Notice that all 5 roots have the same magnitude of 1, so graphically they lie on a circle of radius 1. Also, they all differ by $\frac{360°}{5} = 72°$, so they spaced out evenly on the circle.

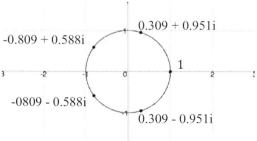

Practice 17

Plot each complex number and convert it to polar form.

1) $5 - 5\sqrt{3}i$
2) $-4i$
3) 3
4) $1 - i$

Find $z_1 \times z_2$ and $z_1 \div z_2$. Leave your answers in polar form.

5) $z_1 = 3(\cos\frac{3\pi}{8} + i\sin\frac{3\pi}{8})$, $z_2 = 7(\cos\frac{\pi}{2} + i\sin\frac{\pi}{2})$

6) $z_1 = (\cos\frac{\pi}{6} + i\sin\frac{\pi}{6})$, $z_2 = 4(\cos\frac{\pi}{10} + i\sin\frac{\pi}{10})$

7) $z_1 = 5(\cos110° + i\sin110°)$, $z_2 = 6(\cos80° + i\sin80°)$

8) $z_1 = \sqrt{3}(\cos20° + i\sin20°)$, $z_2 = \sqrt{11}(\cos25° + i\sin25°)$

Simplify. Leave your answers in standard form.

9) $[\sqrt{2}(\cos\frac{2\pi}{3} + i\sin\frac{2\pi}{3})]^{12}$

10) $[\frac{1}{2}(\cos15° + i\sin15°)]^6$

11) $(\sqrt{3} + i)^8$

12) $(3 - 4i)^5$

13) $(-12 + 5i)^3$

14) $(1 + 2i)^4$

Find the indicated roots of the complex number.

15) Cube roots of $54(\cos210° + i\sin210°)$

16) Fifth roots of $32(\cos300° + i\sin300°)$

17) Fourth roots of $20(-3 + 3i)$

18) Six roots of -2

19) Fifth roots of $4i$

20) Fourth roots of $(4 + 3i)$

Bonus: Solve the equation $x^6 + 2x^3 + 5 = 0$.

Lesson 18: Introduction to Vectors

A **vector** is a quantity that has both <u>magnitude</u> and <u>direction</u>. For example, the <u>velocity</u> of an airplane that travels 200 miles per hour in the North-East direction is a vector quantity. You can represent this quantity using the combined idea of a ray and a segment. A vector is a ray in the sense that it has a direction, but unlike in Geometry, this ray does not go on forever in one direction. A vector has finite magnitude or length; so it is a segment in that sense. Naturally, two vectors that have the same length and direction are equal. Here are some equivalent vectors.

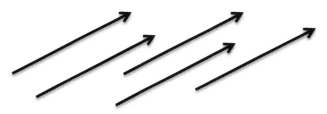

As before, we refer to the starting position as the **initial point** and the final position as the **terminal point**.

As you will see, there are four different ways to describe a vector. One way is you simply tell people the initial and terminal points of your vector. For example, you can say vector **v** start from $P = (2, 1)$ to $Q = (5, 5)$. However, if we all agree to always make our vectors start at $(0, 0)$, then we only need to specify the terminal points. The question is, how do we find the vector that starts at $(0, 0)$ and is equivalent to **v**? Well, it turns out that the solution is quite simple.

To get the new terminal point, you subtract the x- and y-components of P from Q. So, if we let **u** be a vector with initial point $(0, 0)$ and terminal point $(5 - 2, 5 - 1)$ $= (3, 4)$, then **u** = **v**. Here is the figure.

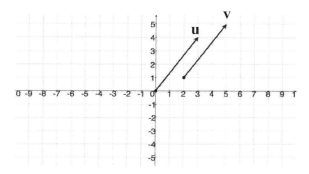

Note that you can use the distance formula to verify that both **u** and **v** have the same magnitude. Also, you can check their slopes to see that they are both equal and therefore are parallel. The advantage of always having the initial point at $(0, 0)$ is that we can then forget about it. So, to describe vector **u** above, we can just say **u** = <3, 4>; this is called the **component form** of vector **u**. Notice that instead of using parentheses, we use "<," and ">" to remind you that "3, 4" is not a single point. When you see <3, 4>, it means the vector consists of two points; namely, $(0, 0)$ and $(3, 4)$.

If **w** is a vector such that **w** = <0, 0>, then it is called the **zero vector**, or **0**. The zero vector has 0 magnitude; so it is just a dot at the origin.

In general, to convert a vector **v** with initial point $P = (p_1, p_2)$ and terminal point $Q = (q_1, q_2)$ into component form, use the following formula:

$$\mathbf{v} = <q_1 - p_1, q_2 - p_2> = <v_1, v_2>$$

Example 1:   Given vector **v** with initial point $(3, 2)$ and terminal point $(-4, -6)$. Convert **v** into component form and find its magnitude.

Solution: $\mathbf{v} = <q_1 - p_1, q_2 - p_2> = <-4 - 3, -6 - 2> = <-7, -8>$.

The magnitude of **v** is $\|\mathbf{v}\| = \sqrt{(-7)^2 + (-8)^2} = \sqrt{113}$.

Now suppose we have two vectors $\mathbf{u} = <u_1, u_2>$ and $\mathbf{v} = <v_1, v_2>$.

1) We can add them and get the vector $\mathbf{u} + \mathbf{v} = <u_1 + v_1, u_2 + v_2>$.

2) We can subtract and get the vector $\mathbf{u} - \mathbf{v} = <u_1 - v_1, u_2 - v_2>$.

3) We can multiply a vector by a real number or scalar k and get the vector
   $k\mathbf{v} = <kv_1, kv_2>$.

Based on this definition of vector addition, you can conclude the following:

1) $\mathbf{u} + \mathbf{v} = \mathbf{v} + \mathbf{u}$.

2) $\mathbf{u} + (\mathbf{v} + \mathbf{w}) = (\mathbf{u} + \mathbf{v}) + \mathbf{w}$.

3) $\mathbf{u} + \mathbf{0} = \mathbf{u}$

4) $\mathbf{u} + (-\mathbf{u}) = \mathbf{0}$

5) $a(b\mathbf{u}) = (ab)\mathbf{u}$

6) $(a + b)\mathbf{u} = a\mathbf{u} + b\mathbf{u}$

7) $a(\mathbf{u} + \mathbf{v}) = a\mathbf{u} + a\mathbf{v}$

8) $0(\mathbf{u}) = \mathbf{0}$, $1(\mathbf{u}) = \mathbf{u}$

9) $\|-\mathbf{u}\| = \|\mathbf{u}\|$

10) $\|c\mathbf{u}\| = |c|\|\mathbf{u}\|$

***Example 2*:**   Given $\mathbf{v} = <2, 5>$ and $\mathbf{w} = <-1, 3>$.

Find a) $3\mathbf{w}$ b) $2\mathbf{v} + \mathbf{w}$ c) $\mathbf{v} - 2\mathbf{w}$

Solution: a) $3\mathbf{w} = 3<-1, 3> = <3(-1), 3(3)> = <-3, 9>$.

b) $2\mathbf{v} + \mathbf{w} = <2(2), 2(5)> + <-1, 3> = <3, 13>$.

c) $\mathbf{v} - 2\mathbf{w} = <2, 5> - <2(-1), 2(3)> = <4, -1>$.

Graphically, you can use the **triangle law** to find the sum $\mathbf{u} + \mathbf{v}$ of two vectors as follows: 1) Move $\mathbf{v}$ so that its initial point touches the terminal point of $\mathbf{u}$, make sure not to change the magnitude and direction in the process. 2) Draw a ray from the initial point of $\mathbf{u}$ to terminal point of $\mathbf{v}$ to form a triangle; this new ray is $\mathbf{u} + \mathbf{v}$. Here is the figure.

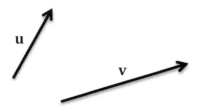
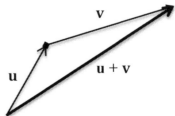

To find **u** - **v**, use the above method and draw **u** + (-**v**), where -**v** is the same as **v,** but the arrow points in the opposite direction.

To find k**v**, draw a vector in the same direction as **v** but k times its magnitude.

Example 3:   Given vectors **v** and **w** as shown below, find $\frac{1}{2}$**v** + 2**w**.

Solution: First draw $\frac{1}{2}$**v** and 2**w**. Then use the triangle law to add them.

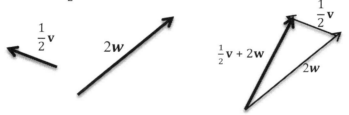

Next, I will talk about unit vectors. A **unit vector** is a vector of length 1.

Given a vector **v**, to find a unit vector **u** that has the same direction as **v**, divide **v** by its magnitude. That is, $\mathbf{u} = \dfrac{\mathbf{v}}{\|\mathbf{v}\|}$

Example 4: Find a unit vector in the same direction as $\mathbf{v} = <-6, 5>$.

Solution: First find the magnitude of $\mathbf{v}$. $\|\mathbf{v}\| = \sqrt{(-6)^2 + (5)^2} = \sqrt{61}$

$$\mathbf{u} = \frac{\mathbf{v}}{\|\mathbf{v}\|} = \frac{1}{\sqrt{61}} \cdot <-6, 5> = <-\frac{6}{\sqrt{61}}, \frac{5}{\sqrt{61}}>.$$

You can check that $\mathbf{u}$ is a unit vector by computing its magnitude.

$$\|\mathbf{u}\| = \sqrt{\left(\frac{-6}{\sqrt{61}}\right)^2 + \left(\frac{5}{\sqrt{61}}\right)^2} = \sqrt{\frac{36}{61} + \frac{25}{61}} = \sqrt{\frac{61}{61}} = 1 \quad \checkmark$$

Here is the figure showing $\mathbf{u}$ and $\mathbf{v}$.

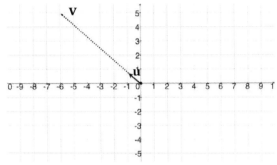

Out of all possible unit vectors, there are two special unit vectors, $\mathbf{i} = <1, 0>$ and $\mathbf{j} = <0, 1>$ these are known as the **standard unit vectors**. Here they are.

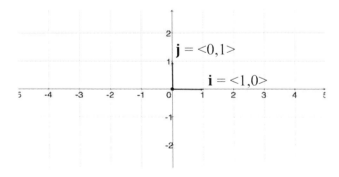

Any vector $\mathbf{v} = <v_1, v_2>$ can be written as a linear combination of $\mathbf{i}$ and $\mathbf{j}$ as follows: $\mathbf{v} = <v_1, v_2> = v_1<1, 0> + v_2<0, 1> = v_1\mathbf{i} + v_2\mathbf{j}$. This is the third way to represent vectors. We call v_1 and v_2 the **horizontal** and **vertical** components of $\mathbf{v}$, respectively.

Example 5: Given **v** = 4**i** - 7**j** and **w** = -5**i** - 2**j**. Find 3**v** - 6**w**.

Solution: 3**v** - 6**w** = 3(4**i** - 7**j**) - 6(-5**i** - 2**j**)
$\qquad\qquad$ = 12**i** - 21**j** + 30**i** + 12**j**
$\qquad\qquad$ = 42**i** - 9**j**

$\qquad$ Note that if you convert **v** and **w** into component form first, you
$\qquad$ will get the same answer.

Now I will talk about the fourth way to represent vectors. Given **v** = <v_1, v_2>,
and let θ be the angle that vector **v** makes with the x-axis.

$\qquad\qquad\qquad\qquad$ From the diagram, we have

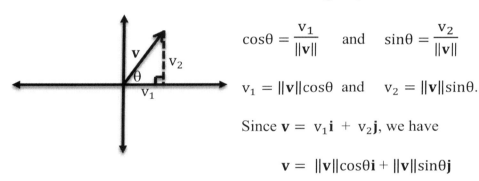

$$\cos\theta = \frac{v_1}{\|\mathbf{v}\|} \quad \text{and} \quad \sin\theta = \frac{v_2}{\|\mathbf{v}\|}$$

$v_1 = \|\mathbf{v}\|\cos\theta$ and $\quad v_2 = \|\mathbf{v}\|\sin\theta$.

Since **v** = v_1**i** + v_2**j**, we have

$$\mathbf{v} = \|\mathbf{v}\|\cos\theta\mathbf{i} + \|\mathbf{v}\|\sin\theta\mathbf{j}$$

Example 6: Given vector **v** with initial point (1, 2) and terminal point (5, -1).
$\qquad\qquad$ Convert **v** into the other three forms.

Solution: In component form, **v** = <5 - 1, -1 - 2> = $\boxed{<4, -3>}$

$\qquad$ This also means $\boxed{\mathbf{v} = 4\mathbf{i} - 3\mathbf{j}}$

$\qquad$ In addition, $\|\mathbf{v}\| = \sqrt{4^2 + (-3)^2} = 5$, and

$$tan\theta = \frac{-3}{4} \implies \theta = \text{-36.87°}.$$

$\qquad$ So, **v** = $\|\mathbf{v}\|\cos\theta\mathbf{i} + \|\mathbf{v}\|\sin\theta\mathbf{j}$

$\qquad$ = $\boxed{5\cos(\text{-36.87°})\mathbf{i} + 5\sin(\text{-36.87°})\mathbf{j}}$

Practice 18

Given vector **v** with the initial point P and terminal point Q, write it in component form and find its magnitude.

1) P = (-2, 4), Q = (1, -4) 　　　　　　2) P = (0, 5), Q = (5, 0)

3) P = (-6, -3), Q = (0, 0) 　　　　　　4) P = (-1, -2), Q = (-7, -2)

Find a unit vector that has the same direction as vector **v**.

5) **v** = <1, 2> 　　　　　　　　　　6) **v** = <-9, 0>

7) **v** = 15**i** - 15**j** 　　　　　　　　8) **v** = -8**i** + 6**j**

Given vectors **u** and **v**, find 2**u** + **v** and **u** - 3**v**.

9) **u** = **i** - **j** and **v** = -2**i** + 3**j** 　　　10) **u** = 4**j** and **v** = 2**i** + 6**j**

11) **u** = <7, -5> and **v** = <-1, -2> 　　12) **u** = <1, 0> and **v** = <0, -1>

Find vector **v** with the given magnitude and has the same direction as vector **w**.

13) $\|\mathbf{v}\|$ = 10, **w** = <-9, 12> 　　　14) $\|\mathbf{v}\|$ = 4, **w** = -**i** + 2**j**

15) $\|\mathbf{v}\|$ = 2, **w** = $3\cos\frac{\pi}{6}\mathbf{i} + 2\sin\frac{\pi}{6}\mathbf{j}$ 　　16) $\|\mathbf{v}\|$ = 3, **w** = $\cos\frac{\pi}{4}\mathbf{i} + \sin\frac{\pi}{4}\mathbf{j}$

Find the angle that vector **v** makes with the positive x-axis.

17) **v** = -3**i** + **j** 　　　　　　　　18) **v** = 9**i** - 15**j**

Find the sum **v** + **w** and the difference **v** - **w** graphically.

19) 　　　　　　　　　　　　　20)

Lesson 19: Dot Products

In the previous lesson, you learned how to add and subtract vectors, as well as multiply a vector by a scalar. In this lesson, you will learn how to multiply two vectors.

Here is the definition of the **dot product** of two vectors.

Given $\mathbf{u} = <u_1, u_2>$ and $\mathbf{v} = <v_1, v_2>$, the dot product is

$$\boxed{\mathbf{u} \cdot \mathbf{v} = u_1 v_1 + u_2 v_2}$$

Note that the answer to the dot product is a scalar, not a vector.

Example 1:   Let $\mathbf{u} = <4, -6>$ and $\mathbf{v} = <-3, 2>$.

Find: a) $\mathbf{u} \cdot \mathbf{v}$ b) $\mathbf{v} \cdot \mathbf{u}$ c) $\mathbf{u} \cdot \mathbf{u}$ d) $\mathbf{v} \cdot \mathbf{v}$

Solution: a) $\mathbf{u} \cdot \mathbf{v} = <4, -6> \cdot <-3, 2> = 4(-3) + (-6)(2) = -24$

b) $\mathbf{v} \cdot \mathbf{u} = <-3, 2> \cdot <4, -6> = (-3)(4) + (2)(-6) = -24$

c) $\mathbf{u} \cdot \mathbf{u} = <4, -6> \cdot <4, -6> = 4(4) + (-6)(-6) = 52$

d) $\mathbf{v} \cdot \mathbf{v} = <-3, 2> \cdot <-3, 2> = (-3)(-3) + (2)(2) = 13$

Notice that $\mathbf{u} \cdot \mathbf{v} = \mathbf{v} \cdot \mathbf{u}$. In general, given vectors $\mathbf{u}$, $\mathbf{v}$, and $\mathbf{w}$, here are the properties of the dot product.

1) $\mathbf{u} \cdot \mathbf{v} = \mathbf{v} \cdot \mathbf{u}$
2) $\mathbf{u} \cdot (\mathbf{v} + \mathbf{w}) = \mathbf{u} \cdot \mathbf{v} + \mathbf{u} \cdot \mathbf{w}$
3) $\mathbf{v} \cdot \mathbf{v} = \|\mathbf{v}\|^2$
4) $\mathbf{0} \cdot \mathbf{v} = 0$

Example 2:   Given $\mathbf{u}$ = <2, -1>, $\mathbf{v}$ = <-3, -5>, and $\mathbf{w}$ = <3, 1>.

Find: a) $(\mathbf{v} \cdot \mathbf{w})\mathbf{u}$ b) $(\mathbf{v} - \mathbf{w}) \cdot \mathbf{u}$ c) $2\mathbf{v} \cdot 4\mathbf{w}$

Solution: a) $(\mathbf{v} \cdot \mathbf{w})\mathbf{u}$ = (<-3, -5> · <3, 1>) <2, -1>

$= [(-3)(3) + (-5)(1)]$ <2, -1>

$= (-14)$ <2, -1>

$=$ <-28, 14>

b) $(\mathbf{v} - \mathbf{w}) \cdot \mathbf{u}$ = (<-3, -5> - <3, 1>) · <2, -1>

$=$ (<-6, -6>) · <2, -1>

$= (-6)(2) + (-6)(-1) = -6$

c) $2\mathbf{v} \cdot 4\mathbf{w}$ = 2 <-3, -5> · 4 <3, 1>

$=$ <-6, -10> · <12, 4>

$= (-6)(12) + (-10)(4) = -72 + -40 = -112.$

Next consider two vectors $\mathbf{u}$ and $\mathbf{v}$ as shown below. The question is, what is the measure of angle θ?

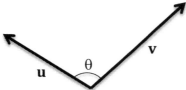

Note that if we add a third side to make a triangle out of the above diagram, then we can use the law of cosines to find θ. But, what is the length of the third side? It turns out that the third side is the vector $-\mathbf{u} + \mathbf{v}$ or $\mathbf{v} - \mathbf{u}$.

Here is the diagram.

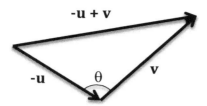

According to the law of cosines, $\|v - u\|^2 = \|u\|^2 + \|v\|^2 - 2\|u\|\|v\|\cos\theta$.

Using the third property of the dot product above, we have
$$(v - u) \cdot (v - u) = u \cdot u + v \cdot v - 2\|u\|\|v\|\cos\theta$$

$$\Rightarrow \quad v \cdot v - v \cdot u - u \cdot v + u \cdot u = u \cdot u + v \cdot v - 2\|u\|\|v\|\cos\theta$$

$$\Rightarrow \quad -v \cdot u - u \cdot v = -2\|u\|\|v\|\cos\theta$$

$$\Rightarrow \quad -2(u \cdot v) = -2\|u\|\|v\|\cos\theta \quad (\text{because } v \cdot u = u \cdot v)$$

$$\Rightarrow \quad u \cdot v = \|u\|\|v\|\cos\theta$$

$$\Rightarrow \quad \cos\theta = \frac{u \cdot v}{\|u\|\|v\|} \quad \text{or} \quad \boxed{\theta = \cos^{-1}\left(\frac{u \cdot v}{\|u\|\|v\|}\right)}$$

Example 3:   Find the angle θ between $u = <5, 1>$ and $v = <2, 7>$.

Solution: First find $u \cdot v$, $\|u\|$, and $\|v\|$.
$$u \cdot v = (5)(2) + (1)(7) = 17$$
$$\|u\| = \sqrt{5^2 + 1^2} = \sqrt{26}$$
$$\|v\| = \sqrt{2^2 + 7^2} = \sqrt{53}$$

$$\theta = \cos^{-1}\left(\frac{u \cdot v}{\|u\|\|v\|}\right) = \cos^{-1}\left(\frac{17}{\sqrt{26}\sqrt{53}}\right)$$
$$= \cos^{-1}(0.4580) = 62.7°$$

Note that since $u \cdot v = \|u\|\|v\|\cos\theta$, if $u \cdot v = 0$, then $\|u\|\|v\|\cos\theta = 0$. This implies $\theta = 90°$. So, we can conclude that two vectors u, v are perpendicular or **orthogonal** if $u \cdot v = 0$.

Example 4:   Are $\mathbf{u} = \,<1, -2>$ and $\mathbf{v} = \,<4, 2>$ orthogonal?

Solution: Since $\mathbf{u} \cdot \mathbf{v} = \,<1, -2> \cdot <4, 2> = (1)(4) + (-2)(2) = 0$, yes $\mathbf{u}$ and $\mathbf{v}$ are orthogonal.

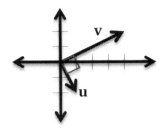

You have now learned how to add two vectors. In physics, sometimes you need to decompose a given vector into the sum of two orthogonal vectors.

For example, suppose you have an object sitting on an inclined as shown.

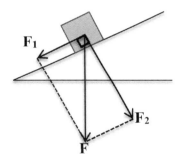

$\mathbf{F}$ is the force of gravity acting on the object, and it pulls the object down as well as pushes the object against the inclined. So, $\mathbf{F} = \mathbf{F_1} + \mathbf{F_2}$. Knowing $\mathbf{F}$, the question is, how do we find the force of friction $\mathbf{F_1}$ and the normal force $\mathbf{F_2}$? (Note: **Normal** means **perpendicular** or **orthogonal**).

Here is another way to look at the same problem. Given vectors $\mathbf{u}$ and $\mathbf{v}$ as shown, how do we find the magnitude of $\mathbf{u_1}$?

Since $\cos\theta = \dfrac{\|\mathbf{u_1}\|}{\|\mathbf{u}\|}$, we have

$$\|\mathbf{u_1}\| = \|\mathbf{u}\|\cos\theta.$$

But $\cos\theta = \dfrac{\mathbf{u} \cdot \mathbf{v}}{\|\mathbf{u}\|\|\mathbf{v}\|},$

so $\|\mathbf{u_1}\| = \|\mathbf{u}\| \left(\dfrac{\mathbf{u} \cdot \mathbf{v}}{\|\mathbf{u}\|\|\mathbf{v}\|}\right)$

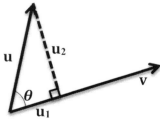

Note: The magnitude of $\mathbf{u_1}$ is also known as the **component of u along v**.

$$\Rightarrow \quad \boxed{\|\mathbf{u_1}\| = \dfrac{\mathbf{u} \cdot \mathbf{v}}{\|\mathbf{v}\|}}$$

Once we know the magnitude of $\mathbf{u_1}$, we can multiply it by the unit vector that has the same direction as vector $\mathbf{v}$, the result is vector $\mathbf{u_1}$.

Hence, $\mathbf{u_1} = \dfrac{\mathbf{u \cdot v}}{\|\mathbf{v}\|}\left(\dfrac{\mathbf{v}}{\|\mathbf{v}\|}\right) = \left(\dfrac{\mathbf{u \cdot v}}{\|\mathbf{v}\|^2}\right)\mathbf{v}$

Vector $\mathbf{u_1}$ is also known as the **projection of u onto v,** written as $\mathbf{proj_v u}$.

So, $\boxed{\mathbf{proj_v u} = \left(\dfrac{\mathbf{u \cdot v}}{\|\mathbf{v}\|^2}\right)\mathbf{v}.}$ Since $\mathbf{u} = \mathbf{u_1} + \mathbf{u_2}$, we also have $\boxed{\mathbf{u_2} = \mathbf{u} - \mathbf{u_1}}$

Example 5: Let $\mathbf{u} = <2, 6>$ and $\mathbf{v} = <-4, 3>$.

Find: a) component of $\mathbf{u}$ along $\mathbf{v}$

b) projection of $\mathbf{u}$ onto $\mathbf{v}$

c) decompose $\mathbf{u}$ into $\mathbf{u_1}$ and $\mathbf{u_2}$, where $\mathbf{u_1}$ is parallel to $\mathbf{v}$ and $\mathbf{u_2}$ orthogonal to $\mathbf{v}$.

Solution: a) First find $\mathbf{u} \cdot \mathbf{v}$ and $\|\mathbf{v}\|$.

$\mathbf{u} \cdot \mathbf{v} = <2, 6> \cdot <-4, 3> = 2(-4) + 6(3) = 10$.

$\|\mathbf{v}\| = \sqrt{(-4)^2 + 3^2} = \sqrt{25} = 5$

component of $\mathbf{u}$ along $\mathbf{v}$ is $\dfrac{\mathbf{u \cdot v}}{\|\mathbf{v}\|} = \dfrac{10}{5} = 2$.

b) $\mathbf{proj_v u} = \left(\dfrac{\mathbf{u \cdot v}}{\|\mathbf{v}\|^2}\right)\mathbf{v} = \left(\dfrac{10}{5^2}\right)<-4, 3> <-\dfrac{8}{5}, \dfrac{6}{5}>$.

c) $\mathbf{u_1} = \mathbf{proj_v u} = <-\dfrac{8}{5}, \dfrac{6}{5}>$, and

$\mathbf{u_2} = \mathbf{u} - \mathbf{u_1} = <2, 6> - <-\dfrac{8}{5}, \dfrac{6}{5}> = <\dfrac{18}{5}, \dfrac{24}{5}>$.

Practice 19

Given vectors **u** and **v**, find: a) $\mathbf{u} \cdot \mathbf{v}$ b) $\mathbf{u} \cdot \mathbf{u}$ c) $\mathbf{v} \cdot \mathbf{v}$.

1) $\mathbf{u} = \langle -1, 2 \rangle$, $\mathbf{v} = \langle -8, 12 \rangle$

2) $\mathbf{u} = \langle 4, 0 \rangle$, $\mathbf{v} = \langle -3, -3 \rangle$

3) $\mathbf{u} = 3\mathbf{i} - \mathbf{j}$, $\mathbf{v} = \mathbf{i} + 5\mathbf{j}$

4) $\mathbf{u} = 10\mathbf{i}$, $\mathbf{v} = -\mathbf{j}$

Given vectors **u**, **v**, and **w**, find: a) $(\mathbf{w} \cdot \mathbf{v})\mathbf{u}$ b) $(\mathbf{w} - \mathbf{u}) \cdot \mathbf{v}$ c) $6\mathbf{v} \cdot 3\mathbf{u}$

5) $\mathbf{u} = \langle 7, -5 \rangle$, $\mathbf{v} = \langle 1, -9 \rangle$, $\mathbf{w} = \langle -\sqrt{2}, -\sqrt{2} \rangle$

6) $\mathbf{u} = \langle -\sqrt{3}, 1 \rangle$, $\mathbf{v} = \langle 0, -1 \rangle$, $\mathbf{w} = \langle 2, 3 \rangle$

7) $\mathbf{u} = \mathbf{i} + \mathbf{j}$, $\mathbf{v} = -\mathbf{i} - \mathbf{j}$, $\mathbf{w} = \mathbf{i} - \mathbf{j}$

8) $\mathbf{u} = 2\mathbf{i} + \mathbf{j}$, $\mathbf{v} = -4\mathbf{j}$, $\mathbf{w} = \mathbf{i} - \sqrt{7}\mathbf{j}$

Find the angle θ between **u** and **v**

9) $\mathbf{u} = \langle 4, 5 \rangle$, $\mathbf{v} = \langle -6, 4 \rangle$

10) $\mathbf{u} = \langle -3, 5 \rangle$, $\mathbf{v} = \langle -1, 2 \rangle$

11) $\mathbf{u} = -3\mathbf{i} - 3\mathbf{j}$, $\mathbf{v} = -2\mathbf{i} - 2\mathbf{j}$

12) $\mathbf{u} = 7\mathbf{i} - 4\mathbf{j}$, $\mathbf{v} = \mathbf{i}$

Determine whether **u** and **v** are orthogonal.

13) **u** = <-2, -4>, **v** = <5, -3>

14) **u** = <-6, 1>, **v** = <0, -9>

Given vectors **u** and **v**, find: a) component of **u** along **v**   b) **proj$_v$u**

15) **u** = <2, 6>, **v** = <5, 1>

16) **u** = <-3, 8>, **v** = <-7, 2>

17) **u** = **i** + 7**j**, **v** = 10**i** + 3**j**

18) **u** = 2**i** + 4**j**, **v** = 6**i** - **j**

19) Let **u** = 12**i** + 11**j**, **v** = 16**i** + 5**j** . Decompose **u** into **u$_1$** and **u$_2$**, where **u$_1$** is parallel to **v** and **u$_2$** orthogonal to **v**.

20) A 6500 pounds car is parked on a hill that has a slope of 10°. Find the force required to keep the car from rolling down the hill.

Bonus: Find the tension T₁ and T₂ in each cable shown in the figure.

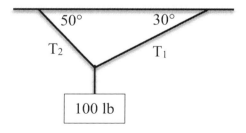

Did you know?

Every year, April 4, June 6, August 8, October 10 and December 12 all fall into the same day of the week as the last day of February.

For odd-numbered months, you can use the "7/11 trick" to remember July 11 and November 7 fall into the same day of the week as the last day of February. Also, since most people go to work from 9 to 5pm, you can use this to remember May 9 and September 5 fall into the same day of the week as the last day of February.

Knowing this, you can quickly tell what day of the week a date falls into. For example, suppose today is Monday February 12, 2018. What day of the week is Christmas?

We know 2/19 and 2/26 are Monday. So, 2/28 is Wednesday. Therefore, 12/12 is Wednesday; so are 12/19 and 12/26. This means 12/25 or Christmas of 2018 falls on Tuesday.

How about July 4, 2018? Well, July 4 is the same as July 11, which is the same as 2/28. So it is Wednesday.

Lesson 20: Vectors In Three Dimension

All of the ideas you learned about two dimensional vectors in the previous two lessons can easily be extended to vectors in space.

Let's start with the three dimensional rectangular coordinate system. Here it is.

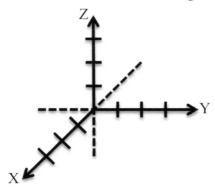

Just as a point in 2-d can be described by an ordered pair (x, y), a point in 3-d can be described by an ordered triple (x, y, z).

Given two points $P = (x_1, y_1, z_1)$ and $Q = (x_2, y_2, z_2)$, the distance from P to Q is

$$d = \sqrt{(x_2 - x_1)^2 + (y_2 - y_1)^2 + (z_2 - z_1)^2}$$

As before, to convert a vector $\mathbf{v}$ with initial point $P = (x_1, y_1, z_1)$ and terminal point $Q = (x_2, y_2, z_2)$ into component form, you can subtract the corresponding components. That is,

$$\mathbf{v} = \langle x_2 - x_1, y_2 - y_1, z_2 - z_1 \rangle.$$

Also, for any nonzero vector $\mathbf{v}$, you can use the previous formula to find a unit vector $\mathbf{u}$ that has the same direction as $\mathbf{v}$. That is, $\mathbf{u} = \dfrac{\mathbf{v}}{\|\mathbf{v}\|}$

Since we are in 3-d, there are now three, instead of two, special unit vectors known as the **standard unit vectors**. Here they are:

$$\mathbf{i} = \langle 1, 0, 0 \rangle, \quad \mathbf{j} = \langle 0, 1, 0 \rangle, \quad \mathbf{k} = \langle 0, 0, 1 \rangle.$$

Example 1: Given vector $\mathbf{v}$ with initial point $(5, 1, -3)$ and terminal point $(4, -2, 7)$. Convert $\mathbf{v}$ into component form and find its magnitude.

Solution: $\mathbf{v} = \langle x_2 - x_1, y_2 - y_1, z_2 - z_1 \rangle = \langle 4 - 5, -2 - 1, 7 - -3 \rangle$
$$= \langle -1, -3, 10 \rangle.$$

The magnitude of $\mathbf{v}$ is $\|\mathbf{v}\| = \sqrt{(-1)^2 + (-3)^2 + (10)^2} = \sqrt{110}.$

Example 2: Let $\mathbf{v} = 2\mathbf{i} + 4\mathbf{j} + 5\mathbf{k}$ and $\mathbf{w} = 3\mathbf{i} - 2\mathbf{j} - \mathbf{k}$. Find $11\mathbf{v} - 8\mathbf{w}$.

Solution: $11\mathbf{v} - 8\mathbf{w} = 11(2\mathbf{i} + 4\mathbf{j} + 5\mathbf{k}) - 8(3\mathbf{i} - 2\mathbf{j} - \mathbf{k})$

$$= 22\mathbf{i} + 44\mathbf{j} + 55\mathbf{k} - 24\mathbf{i} + 16\mathbf{j} + 8\mathbf{k}$$

$$= -2\mathbf{i} + 60\mathbf{j} + 63\mathbf{k}.$$

The definition of dot product for 3-d vectors is again similar to that of the vectors in 2-d. Given $\mathbf{u} = <u_1, u_2, u_3>$ and $\mathbf{v} = <v_1, v_2, v_3>$,

$$\boxed{\mathbf{u} \cdot \mathbf{v} = u_1 v_1 + u_2 v_2 + u_3 v_3}$$

All of the properties of the dot product still applied.

Example 3:   Let $\mathbf{u} = <1, 0, 4>$ and $\mathbf{v} = <5, -2, 2>$.
Find: a) $\mathbf{u} \cdot \mathbf{v}$ b) $\mathbf{v} \cdot \mathbf{u}$

Solution: a) $\mathbf{u} \cdot \mathbf{v} = <1, 0, 4> \cdot <5, -2, 2>$
$$= (1)(5) + (0)(-2) + (4)(2) = 13$$

b) $\mathbf{v} \cdot \mathbf{u} = <5, -2, 2> \cdot <1, 0, 4>$
$$= (5)(1) + (-2)(0) + (2)(4) = 13$$

As you can see, $\mathbf{u} \cdot \mathbf{v} = \mathbf{v} \cdot \mathbf{u}$.

To find the angle θ between two vectors $\mathbf{u}$ and $\mathbf{v}$ in 3-d, you can use the same formula as in 2-d. That is, $\cos\theta = \dfrac{\mathbf{u} \cdot \mathbf{v}}{\|\mathbf{u}\|\|\mathbf{v}\|}$.

Example 4:   Find the angle θ between $\mathbf{u} = <6, 8, 2>$ and $\mathbf{v} = <1, 5, 4>$.

Solution: First find $\mathbf{u} \cdot \mathbf{v}$, $\|\mathbf{u}\|$, and $\|\mathbf{v}\|$.

$$\mathbf{u} \cdot \mathbf{v} = (6)(1) + (8)(5) + (2)(4) = 54$$

$$\|\mathbf{u}\| = \sqrt{6^2 + 8^2 + 2^2} = \sqrt{104}.$$

$$\|\mathbf{v}\| = \sqrt{1^2 + 5^2 + 4^2} = \sqrt{42}.$$

$$\theta = \cos^{-1}\left(\frac{\mathbf{u}\cdot\mathbf{v}}{\|\mathbf{u}\|\|\mathbf{v}\|}\right) = \cos^{-1}\left(\frac{54}{\sqrt{104}\sqrt{42}}\right)$$

$$= \cos^{-1}(0.8171) = 35.2°.$$

Recall that one of the ways to represent a vector in 2-d is to specify its magnitude and the angle it makes with the x-axis. That is, $\mathbf{v} = \|\mathbf{v}\|\cos\theta\mathbf{i} + \|\mathbf{v}\|\sin\theta\mathbf{j}$.

For vectors in 3-d, it is more complicated. Not only you need the magnitude, you also need the three **direction angles** A, B, and C.

    1) A is the angle vector $\mathbf{v}$ makes with $\mathbf{i}$ or the positive x-axis.

    2) B is the angle vector $\mathbf{v}$ makes with $\mathbf{j}$ or the positive y-axis.

    3) C is the angle vector $\mathbf{v}$ makes with $\mathbf{k}$ or the positive z-axis.

Here is the diagram.

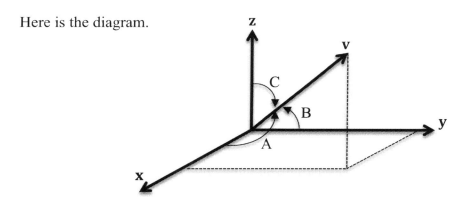

Using the previous formula to find the angle between two vectors, we have

$$\cos A = \frac{\mathbf{v}\cdot\mathbf{i}}{\|\mathbf{v}\|\|\mathbf{i}\|} = \frac{<v_1,v_2,v_3>\cdot<1,0,0>}{\|\mathbf{v}\|\cdot 1} = \frac{v_1}{\|\mathbf{v}\|}$$

$$\cos B = \frac{\mathbf{v}\cdot\mathbf{j}}{\|\mathbf{v}\|\|\mathbf{j}\|} = \frac{<v_1,v_2,v_3>\cdot<0,1,0>}{\|\mathbf{v}\|\cdot 1} = \frac{v_2}{\|\mathbf{v}\|}$$

$$\cos C = \frac{\mathbf{v} \cdot \mathbf{k}}{\|\mathbf{v}\|\|\mathbf{k}\|} = \frac{<v_1, v_2, v_3> \cdot <0,0,1>}{\|\mathbf{v}\| \cdot 1} = \frac{v_3}{\|\mathbf{v}\|}$$

The values of cosA, cosB, and cosC are known as the **direction cosines** of vector **v**. They determine the "steepness" of your vector, just like the "slope" of a line in a plane determine how steep a line is.

Example 5:   Find the direction cosines and direction angles of $\mathbf{v} = <2, 6, 3>$.

Solution: First find the magnitude of **v**.

$$\|\mathbf{v}\| = \sqrt{2^2 + 6^2 + 3^2} = \sqrt{49} = 7.$$

$$\boxed{\cos A = \frac{v_1}{\|\mathbf{v}\|} = \frac{2}{7}} \qquad \boxed{\cos B = \frac{v_2}{\|\mathbf{v}\|} = \frac{6}{7}} \qquad \boxed{\cos C = \frac{v_3}{\|\mathbf{v}\|} = \frac{3}{7}}$$

$$A = \cos^{-1}\left(\frac{2}{7}\right) \qquad B = \cos^{-1}\left(\frac{6}{7}\right) \qquad C = \cos^{-1}\left(\frac{3}{7}\right)$$

$$\boxed{A = 73.4°} \qquad \boxed{B = 31.0°} \qquad \boxed{C = 64.6°}$$

Notice that $\left(\frac{2}{7}\right)^2 + \left(\frac{6}{7}\right)^2 + \left(\frac{3}{7}\right)^2 = \frac{4}{49} + \frac{36}{49} + \frac{9}{49} = 1.$

In general, $\cos^2 A + \cos^2 B + \cos^2 C = 1$.

Before finishing this lesson and the book, there is one more thing that I want to show you, and that is the concept of the **cross product**. Unlike the dot product, the cross product of two vectors is another vector, not a scalar. Also, the cross product only applies to three dimensional vectors. Here is the definition.

Given two vectors $\mathbf{v} = a_1\mathbf{i} + b_1\mathbf{j} + c_1\mathbf{k}$ and $\mathbf{w} = a_2\mathbf{i} + b_2\mathbf{j} + c_2\mathbf{k}$,

$$\boxed{\mathbf{v} \times \mathbf{w} = (b_1 c_2 - b_2 c_1)\mathbf{i} - (a_1 c_2 - a_2 c_1)\mathbf{j} + (a_1 b_2 - a_2 b_1)\mathbf{k}}$$

If you think this formula looks random and hard to memorize, don't worry. There are patterns. Here is how you can remember this formula.

For the **i** or first component, the first letter or "a" is missing. We only have b and c; the subscript is "1, 2 and then 2, 1."

Similarly, for the **j** or second component, the second letter or "b" is missing. We only have a and c; the subscript is "1, 2 and then 2, 1."

Finally, for the **k** or third component, the third letter or "c" is missing. We only have a and b; the subscript is "1, 2 and then 2, 1."

Inside the parentheses, all signs are minuses; outside the parentheses, it is alternating between "positive negative positive."

That is all. With a little practice, you should be able to quickly write down the formula for the cross product. Now, if you studied matrices before, you can also use **determinant** to find the cross product. For this book, I will just use the above formula.

Example 6:   Let $\mathbf{v} = 4\mathbf{i} + 6\mathbf{j} + 2\mathbf{k}$, $\mathbf{w} = \mathbf{i} + 5\mathbf{j} + 6\mathbf{k}$.
    Find: a) $\mathbf{v} \times \mathbf{w}$ b) $\mathbf{w} \times \mathbf{v}$ c) $\mathbf{v} \times \mathbf{v}$

Solution: a) $a_1 = 4, b_1 = 6, c_1 = 2$; $a_2 = 1, b_2 = 5, c_2 = 6$

$$\mathbf{v} \times \mathbf{w} = (b_1 c_2 - b_2 c_1)\mathbf{i} - (a_1 c_2 - a_2 c_1)\mathbf{j} + (a_1 b_2 - a_2 b_1)\mathbf{k}$$

$$= (6 \cdot 6 - 5 \cdot 2)\mathbf{i} - (4 \cdot 6 - 1 \cdot 2)\mathbf{j} + (4 \cdot 5 - 1 \cdot 6)\mathbf{k}$$

$$= 26\mathbf{i} - 22\mathbf{j} + 14\mathbf{k}$$

b) $a_1 = 1, b_1 = 5, c_1 = 6$; $a_2 = 4, b_2 = 6, c_2 = 2$

$$\mathbf{w} \times \mathbf{v} = (b_1 c_2 - b_2 c_1)\mathbf{i} - (a_1 c_2 - a_2 c_1)\mathbf{j} + (a_1 b_2 - a_2 b_1)\mathbf{k}$$

$$= (5 \cdot 2 - 6 \cdot 6)\mathbf{i} - (1 \cdot 2 - 4 \cdot 6)\mathbf{j} + (1 \cdot 6 - 4 \cdot 5)\mathbf{k}$$

$$= -26\mathbf{i} + 22\mathbf{j} - 14\mathbf{k}.$$

c) $a_1 = 4$, $b_1 = 6$, $c_1 = 2$; $a_2 = 4$, $b_2 = 6$, $c_2 = 2$

$$\mathbf{v} \times \mathbf{v} = (b_1 c_2 - b_2 c_1)\mathbf{i} - (a_1 c_2 - a_2 c_1)\mathbf{j} + (a_1 b_2 - a_2 b_1)\mathbf{k}$$

$$= (6 \cdot 2 - 6 \cdot 2)\mathbf{i} - (4 \cdot 2 - 4 \cdot 2)\mathbf{j} + (4 \cdot 6 - 4 \cdot 6)\mathbf{k}$$

$$= 0\mathbf{i} - 0\mathbf{j} + 0\mathbf{k} = \mathbf{0}.$$

Notice that $\mathbf{v} \times \mathbf{w}$ does not equal $\mathbf{w} \times \mathbf{v}$. In general, $\mathbf{v} \times \mathbf{w} = -(\mathbf{w} \times \mathbf{v})$.
Also, any vector crosses with itself equal $\mathbf{0}$.
In fact, $\|\mathbf{v} \times \mathbf{w}\| = \|\mathbf{v}\|\|\mathbf{w}\|\sin\theta$ (compare this to $\mathbf{v} \cdot \mathbf{w} = \|\mathbf{v}\|\|\mathbf{w}\|\cos\theta$).
        This means, $\|\mathbf{v} \times \mathbf{w}\| = 0$ if $\theta = 0°$.
So, any vector crosses with a vector that is parallel to itself, the answer is $\mathbf{0}$.
Here are other properties of the cross product that you should know.
1) the vector $\mathbf{v} \times \mathbf{w}$ is orthogonal to both $\mathbf{v}$ and $\mathbf{w}$.
2) $\|\mathbf{v} \times \mathbf{w}\|$ is the area of the parallelogram having $\mathbf{v}$ and $\mathbf{w}$ as adjacent sides.
3) $k(\mathbf{v} \times \mathbf{w}) = (k\mathbf{v}) \times \mathbf{w} = \mathbf{v} \times (k\mathbf{w})$.
4) $\mathbf{u} \times (\mathbf{v} + \mathbf{w}) = (\mathbf{u} \times \mathbf{v}) + (\mathbf{u} \times \mathbf{w})$.

Example 7:   Find a unit vector that is orthogonal to $\mathbf{v} = <-1, 5, -2>$ and
        $\mathbf{w} = <4, 0, -6>$.

<u>Solution</u>:   First find $\mathbf{v} \times \mathbf{w}$, then turn it into a unit vector.
        $\mathbf{v} \times \mathbf{w} = [5 \cdot (-6) - 0 \cdot (-2)]\mathbf{i} - [-1 \cdot (-6) - 4 \cdot (-2)]\mathbf{j} +$
                    $(-1 \cdot 0 - 4 \cdot 5)\mathbf{k} = -30\mathbf{i} - 14\mathbf{j} - 20\mathbf{k}.$

        $\|\mathbf{v} \times \mathbf{w}\| = \sqrt{1496}.$

        The unit vector is $\dfrac{-30}{\sqrt{1496}}\mathbf{i} - \dfrac{14}{\sqrt{1496}}\mathbf{j} - \dfrac{20}{\sqrt{1496}}\mathbf{k}.$

Practice 20

Given vectors **u** and **v**, find: a) **u · v**   b) **u · u**   c) **v · v**.

1) **u** = <8, 1, 0>, **v** = <2, -1, 9>

2) **u** = -12**i**, **v** = 15**k**

Given vectors **v**, and **w**, find: a) **v** × **w**   b) **w** × **v**   c) **v** × **v**

3) **v** = <1, 5, -6>, **w** = <-2, -1, -4>

4) **v** = <0, 0, -1>, **w** = <7, 11, 0>

5) **v** = -2**i** - 2**j** + 2**k**, **w** = **i** + **j** + **k**

6) **v** = -5**j**, **w** = **i** - 20**k**

Find the angle θ between **u** and **v**

7) **u** = <4, 3, 5>, **v** = <-1, 1, -1>

8) **u** = <1, 2, 1>, **v** = <-1, 2, -1>

9) **u** = 2**i** - 4**j** - 6**k**, **v** = 3**i** + 4**j**

10) **u** = -3**j**, **v** = 6**k**

Find the direction cosines and direction angles of the given vector.

11) **v** = <4, 2, 6>

12) **v** = <-5, -1, -8>

13) $\mathbf{v} = 10\mathbf{i} - 20\mathbf{j} + 10\mathbf{k}$

14) $\mathbf{v} = -6\mathbf{i} + 8\mathbf{j} + 10\mathbf{k}$

Find a unit vector that is orthogonal to $\mathbf{u}$ and $\mathbf{v}$.

15) $\mathbf{u} = <0, -2, 4>, \mathbf{v} = <10, 2, -3>$

16) $\mathbf{u} = <5, 3, 6>, \mathbf{v} = <-2, 11, 1>$

17) $\mathbf{u} = 9\mathbf{i} + \mathbf{j} - 2\mathbf{k}, \mathbf{v} = -2\mathbf{i} + 5\mathbf{j} + 3\mathbf{k}$

18) $\mathbf{u} = -7\mathbf{i} - 3\mathbf{j}, \mathbf{v} = 6\mathbf{j} + 5\mathbf{k}$

Given vectors $\mathbf{u}$, $\mathbf{v}$, and $\mathbf{w}$, find $\mathbf{u} \times (\mathbf{v} + \mathbf{w})$.

19) $\mathbf{u} = <-2, 5, 4>, \mathbf{v} = <3, 1, -2>, \mathbf{w} = <-6, 10, -1>$

20) $\mathbf{u} = 6\mathbf{j} - 7\mathbf{k}, \mathbf{v} = 2\mathbf{i} - \mathbf{j} + 4\mathbf{k}, \mathbf{w} = -\mathbf{i} - \mathbf{j} - 2\mathbf{k}$.

Answers for Even Problems

Practice 1

2. $\sin\angle X = \frac{3}{7}$, $\csc\angle X = \frac{7}{3}$ $\cos\angle X = \frac{2\sqrt{10}}{7}$, $\sec\angle X = \frac{7\sqrt{10}}{20}$, $\tan\angle X = \frac{3\sqrt{10}}{20}$, $\cot\angle X = \frac{2\sqrt{10}}{3}$

4. $\sin\angle X = \frac{\sqrt{2}}{2}$, $\csc\angle X = \sqrt{2}$ $\cos\angle X = \frac{\sqrt{2}}{2}$, $\sec\angle X = \sqrt{2}$, $\tan\angle X = 1$, $\cot\angle X = 1$

6. $m\angle A = 50°$, BC=9.53, AC=12.45   **8.** $m\angle A = 71°$, AB=3.44, AC=10.58

10. 8.39   **12.** 1.68   **14.** 1.84   **16.** 1   **18.** 9.90   **20.** 48.73

Practice 2

2. $\frac{9\pi}{4}$ **4.** 22.5°   **6.** 510°   **8.** 260°, 620°, −460°, −820°   **10.** $\frac{19\pi}{6}, \frac{31\pi}{6}, \frac{-5\pi}{6}, \frac{-17\pi}{6}$

12. $1.5 + 2\pi \approx 7.78$, $1.5 + 4\pi \approx 14.07$, $1.5 - 2\pi \approx -4.78$, $1.5 - 4\pi \approx -11.07$

14. 53.9517°   **16.** 89°9′   **18.** 120°4′12″   **20.** 0.1885 rad/sec, 6.593m/sec

Practice 3

2. See "table" in lesson   **4.** $5.5\sqrt{3}$ **6.** $-\frac{\sqrt{3}}{2}$ **8.** -1

10. 0   **12.** $\frac{2\sqrt{3}}{3}$ **14.** 1   **16.** -2   **18.** -2   **20.** undefined

Practice 4

2. $1, \pi, \frac{3\pi}{8}, 0$ **4.** $\frac{9}{5}, \frac{24}{7}, \frac{-6}{35}, -3$ **6.** $\frac{3}{4}, 18, 9, 3$

8.

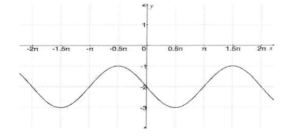

10.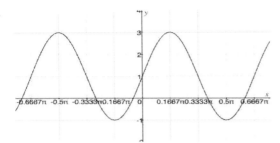

12.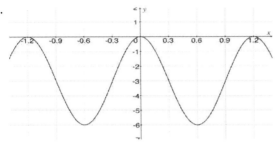

14.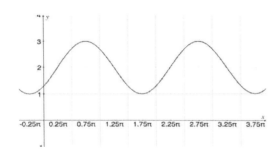

16.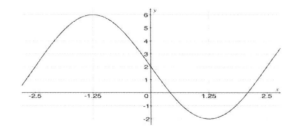

18.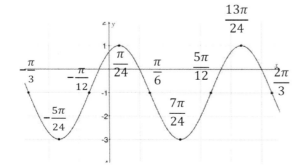

20. 2

Practice 5

2. $\dfrac{\pi}{3}, \dfrac{4\pi}{3}, \dfrac{-2\pi}{3}, \dfrac{-5\pi}{3}$

4. $\dfrac{-\pi}{4}, \dfrac{3\pi}{4}, \dfrac{7\pi}{4}, \dfrac{-5\pi}{4}$

6. $\dfrac{3\pi}{4}$

8. 8

10.

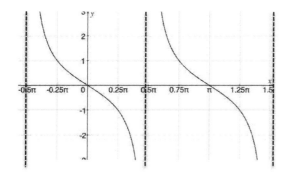

12.

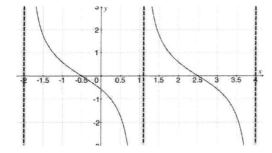

14.

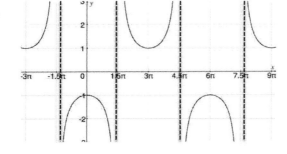

16.

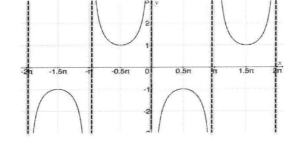

18.

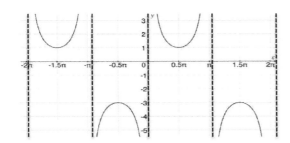

20.

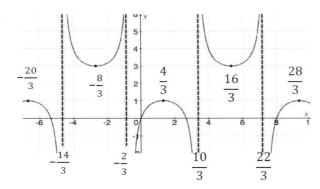

Practice 6

2. $\frac{\pi}{2}$ **4.** $\frac{-\pi}{6}$ or -30°   **6.** 125.3°   **8.** $\frac{-\pi}{6}$ or -30°   **10.** 120°   **12.** $\frac{1}{\sqrt{1-x^2}}$

14.

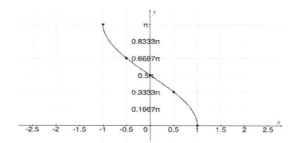

16.

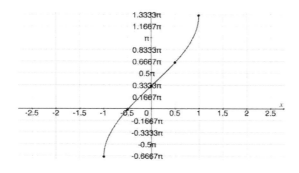

18. -0.67   **20.** 1.25

Practice 7

2. $\sin x = \dfrac{\sqrt{7}}{4}$, $\csc x = \dfrac{4\sqrt{7}}{7}$, $\cos x = \dfrac{3}{4}$, $\sec x = \dfrac{4}{3}$, $\tan x = \dfrac{\sqrt{7}}{3}$, $\cot x = \dfrac{3\sqrt{7}}{7}$

4. $\sin x = \dfrac{-4\sqrt{17}}{17}$, $\csc x = -\dfrac{\sqrt{17}}{4}$, $\cos x = \dfrac{\sqrt{17}}{17}$, $\sec x = \sqrt{17}$, $\tan x = -4$, $\cot x = -\dfrac{1}{4}$

6. $\cos^2 x$ **8.** $3\sin x$ **10.-18.** Answers will vary.   **20.** $12\csc\theta$

Practice 8

2. $x = \dfrac{5\pi}{6} + 2n\pi$, $x = \dfrac{7\pi}{6} + 2n\pi$, where n is an integer.

4. $x = \dfrac{\pi}{3} + n\pi$, $x = \dfrac{2\pi}{3} + n\pi$, where n is an integer.

6. $x = \dfrac{\pi}{4} + n\pi$, $x = \dfrac{3\pi}{4} + n\pi$, where n is an integer.

8. $x = \dfrac{\pi}{3} + n\pi$, $x = \dfrac{2\pi}{3} + n\pi$, where n is an integer.   **10.** $x = \dfrac{\pi}{3}$, $x = \dfrac{5\pi}{3}$

12. $x = \dfrac{1}{3}, \dfrac{13}{3}$, $x = \dfrac{11}{3}$ **14.** $x = 0, \pi, 2.138, 5.279$ **16.** $x = 1.11, 2.03, 4.25, 5.18$

18. $x = \dfrac{2\pi}{3}, \dfrac{4\pi}{3}$, $x = \pi$ **20.** $x = 1.36, 2.57, 4.50, 5.71$

Practice 9

2. $\dfrac{\sqrt{2}-\sqrt{6}}{4}$ **4.** $\dfrac{\sqrt{3}-3}{\sqrt{3}+3} = \dfrac{1-\sqrt{3}}{1+\sqrt{3}}$ **6.** $\dfrac{1+\sqrt{3}}{1-\sqrt{3}}$ **8.** $\dfrac{4\sqrt{3}+3}{10}$ **10.** $\csc x$

12. $\dfrac{1+\sqrt{3}\tan x}{\sqrt{3}-\tan x}$ **14.** $\dfrac{72+14\sqrt{10}}{-21+48\sqrt{10}}$ **16.** $\dfrac{175}{-21-48\sqrt{10}}$ **18.** Answer will vary.

20. $\tan(x - y) = \tan[x + (-y)] = \dfrac{\tan x + \tan(-y)}{1 - \tan x \tan(-y)} = \dfrac{\tan x - \tan(y)}{1 + \tan x \tan(y)}$

Practice 10

2. $\dfrac{7}{25}$ **4.** $\dfrac{117}{44}$ **6.** $\dfrac{25}{7}$ **8.** $\sqrt{\dfrac{2+\sqrt{2}}{2-\sqrt{2}}}$ **10.** $-\dfrac{\sqrt{2+\sqrt{2}}}{2}$

12. $\sqrt{\dfrac{2+\sqrt{2}}{2-\sqrt{2}}}$ **14.** $\dfrac{1-\cos4x}{8}$ **16.** Answer will vary.

18. $x = 0.7227 + 2n\pi$, $x = 5.5605 + 2n\pi$, $x = n\pi$ where n is an odd integer.

20. $\dfrac{1}{2x\sqrt{1-x^2}}$

Practice 11

2. $\dfrac{1}{2}\sin6x + \dfrac{1}{2}\sin4x$ **4.** $\dfrac{1}{2}\cos\dfrac{13}{6}x + \dfrac{1}{2}\cos\dfrac{11}{6}x$ **6.** $5\cos y - 5\cos x$

8. $8\cos\dfrac{7}{2}x \cos\dfrac{1}{2}x$ **10.** $-2\sin\dfrac{6}{35}x \sin\dfrac{1}{35}x$ **12.** $\sqrt{3}\sin x$

14.-16. Answers will vary.

18. $x = \dfrac{n\pi}{2}$, where n is an integer. $x = \dfrac{n\pi}{6}$, where n is an odd integer.

20. $x = \dfrac{24n}{7}$, $x = 24n$, where n is an integer.

Practice 12

2. $m\angle C = 30°, a = 34.4, b = 54.4$ **4.** $m\angle B = 68.3°, m\angle C = 63.7°, c = 24.1$
or $m\angle B = 111.7°, m\angle C = 20.3°, c = 9.3$

6. $m\angle B = 23.2°, m\angle C = 123.8°, c = 27.5$ **8.** $m\angle B = 51.5°, m\angle A = 18.5°, a = 24.3$

10. $m\angle C = 38.3°, m\angle B = 39.7°, b = 19.6$

12. 2 **14.** 1 **16.** 2 **18.** 0 **20.** 27.5 ft

Practice 13

2. $m\angle C = 71.8°, m\angle B = 61.3°, m\angle A = 46.9°$ **4.** $m\angle B = 39.7°, m\angle C = 75.3°, a = 8.4$

6. $m\angle A = 37.7°, m\angle B = 102.3°, c = 5.3$

8. $m\angle C = 37.1°, m\angle A = 71.5°, m\angle B = 71.5°$

10. $m\angle A = 27.0°, m\angle B = 62.8°, c = 2.2$

12. $131.4°$ **14.** 86.1 **16.** $60°$ **18.** N75.5°E   **20.** N60.3°W

Practice 14

2. 19.1   **4.** 876.9   **6.** 87.9   **8.** 912.7   **10.** 2.0   **12.** 18.0   **14.** 5.8

16.   small triangle on the left $= \frac{1}{2} \cdot 10 \cdot a \cdot sinx = 5asinx$

small triangle on the right $= \frac{1}{2} \cdot 10 \cdot a \cdot sin(180 - x) = 5asinx$

total area $= 5asinx + 5asinx = 10asinx$

18. 10.6   **20.** 100.2

Practice 15

2.

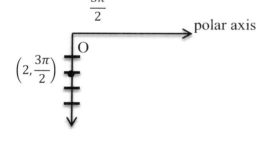

4.

6.

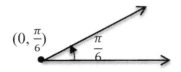

8. $(2, \frac{7\pi}{4}), (-2, \frac{3\pi}{4}), (-2, \frac{-5\pi}{4})$

10. $(1.5, 2 - 2\pi), (-1.5, \pi + 2), (-1.5, -\pi + 2)$ **12.** $(5\sqrt{3}, -5)$

14. $(-1.98, 0.28)$ **16.** $(10, -53.1°)$ **18.** $(\sqrt{5}, 219.2°)$ **20.** $r = tan\theta sec\theta$

Practice 16

2. circle　　　　**4**. cardioid　　　**6**. rose curve　　　　**8**. straight line

10.

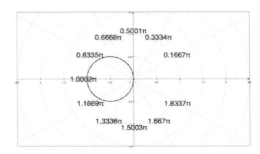

12.

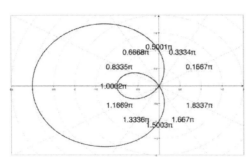

14.

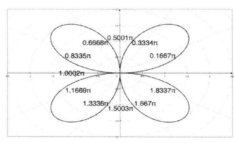

16.

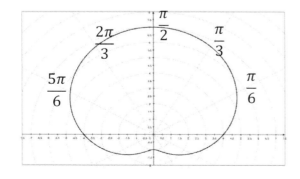

18. $\tan\theta = \tan\frac{\pi}{8} = 0.41 = \frac{y}{x} \Rightarrow y = 0.41x$

20. $y = \frac{1}{8}x^2 - 2$

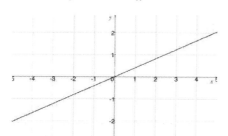

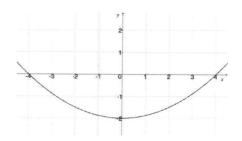

Practice 17

2. $4[cos\left(\frac{-\pi}{2}\right) + isin\left(\frac{-\pi}{2}\right)]$

4. $\sqrt{2}[cos\left(\frac{-\pi}{4}\right) + isin\left(\frac{-\pi}{4}\right)]$

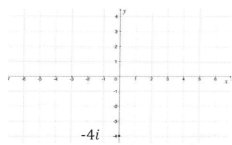

-4i

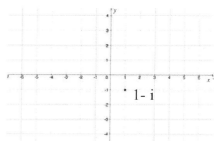

1- i

6. $z_1 \times z_2 = 4(cos\frac{4\pi}{15} + isin\frac{4\pi}{15})$, $z_1 \div z_2 = \frac{1}{4}(cos\frac{\pi}{15} + isin\frac{\pi}{15})$

8. $z_1 \times z_2 = \sqrt{33}(cos45° + isin45°)$, $z_1 \div z_2 = \frac{\sqrt{33}}{11}[cos(-5°) + isin(-5°)]$

10. $\frac{1}{64}i$ **12.** $-245.2 + 3098.3i$ **14.** $-7 - 24i$

16. $1 + \sqrt{3}i, -1.3 + 1.5i, -1.8 + 0.8i, 0.2 - 2.0i, 2.0 - 0.4i$

18. $\sqrt[6]{2}(\frac{\sqrt{3}}{2} + \frac{1}{2}i), \sqrt[6]{2}i, \sqrt[6]{2}(-\frac{\sqrt{3}}{2} + \frac{1}{2}i), \sqrt[6]{2}(-\frac{\sqrt{3}}{2} - \frac{1}{2}i), -\sqrt[6]{2}i, \sqrt[6]{2}(\frac{\sqrt{3}}{2} - \frac{1}{2}i)$

20. $\sqrt[4]{5}(0.987 + 0.160i), \sqrt[4]{5}(-0.160 + 0.987i),$
      $\sqrt[4]{5}(-0.987 - 0.160i), \sqrt[4]{5}(0.160 - 0.987i).$

Practice 18

2. $<5, -5>, 5\sqrt{2}$ **4.** $<-6, 0>, 6$ **6.** $<-1, 0>$ **8.** $\frac{-4}{5}\mathbf{i} + \frac{3}{5}\mathbf{j}$

10. $2\mathbf{i} + 14\mathbf{j}$, $-6\mathbf{i} - 14\mathbf{j}$ **12**. $<2, -1>, <1, 3>$ **14**. $\frac{-4}{\sqrt{5}}\mathbf{i} + \frac{8}{\sqrt{5}}\mathbf{j}$

16. $\frac{3\sqrt{2}}{2}\mathbf{i} + \frac{3\sqrt{2}}{2}\mathbf{j}$ **18**. $-59.0°$ **20**.

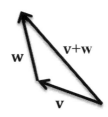
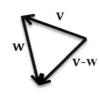

Practice 19

2. $-12, 16, 18$ **4**. $0, 100, 1$ **6**. $<3\sqrt{3}, -3>, -2, -18$

8. $8\sqrt{7}\mathbf{i} + 4\sqrt{7}\mathbf{j}, 14.583 - 144$ **10**. $4.4°$ **12**. $29.7°$ **14**. No

16. $5.082, <-\frac{259}{53}, \frac{74}{53}>$ **18**. $1.315, <\frac{48}{37}, \frac{-8}{37}>$ **20**. 1129 pounds

Practice 20

2. $0, 144, 225$ **4**. $11\mathbf{i} - 7\mathbf{j}, -11\mathbf{i} + 7\mathbf{j}, \mathbf{0}$ **6**. $100\mathbf{i} + 5\mathbf{k}, -100\mathbf{i} - 5\mathbf{k}, \mathbf{0}$

8. $70.5°$ **10**. $90°$

12. $cosA = \frac{-5}{3\sqrt{10}}, cosB = \frac{-1}{3\sqrt{10}}, cosC = \frac{-8}{3\sqrt{10}}, A = 121.8°, B = 96.1°, C = 147.5°$

14. $cosA = \frac{-3}{5\sqrt{2}}, cosB = \frac{4}{5\sqrt{2}}, cosC = \frac{1}{\sqrt{2}}, A = 115.1°, B = 55.6°, C = 45°$

16. $-\frac{63}{\sqrt{7979}}\mathbf{i} - \frac{17}{\sqrt{7979}}\mathbf{j} + \frac{61}{\sqrt{7979}}\mathbf{k}$ **18**. $-\frac{15}{\sqrt{3214}}\mathbf{i} + \frac{35}{\sqrt{3214}}\mathbf{j} - \frac{42}{\sqrt{3214}}\mathbf{k}$

20. $-2\mathbf{i} + 7\mathbf{j} - 6\mathbf{k}$